Neuropedia

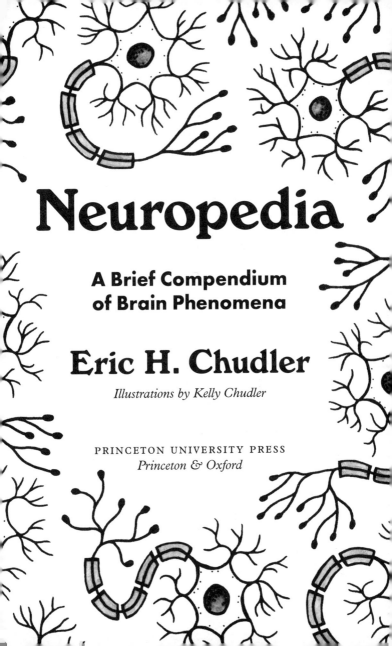

Neuropedia

A Brief Compendium of Brain Phenomena

Eric H. Chudler

Illustrations by Kelly Chudler

PRINCETON UNIVERSITY PRESS
Princeton & Oxford

Published by Princeton University Press
41 William Street, Princeton, New Jersey 08540
99 Banbury Road, Oxford OX2 6JX

press.princeton.edu

All Rights Reserved

ISBN 978-0-691-21357-6
ISBN (e-book) 978-0-691-24218-7

British Library Cataloging-in-Publication Data is available

Editorial: Hallie Stebbins and Kiran Pandey
Production Editorial: Mark Bellis
Text and Cover Design: Chris Ferrante
Production: Steve Sears
Publicity: Sara Henning-Stout and Kate Farquhar-Thomson
Copyeditor: Jennifer McClain

Cover, endpaper, and text illustrations by Kelly Chudler

This book has been composed in Plantin, Futura, and Windsor

Printed on acid-free paper. ∞

Printed in China

10 9 8 7 6 5 4 3 2 1

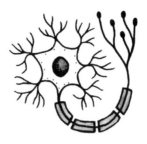

Preface

Mientras nuestro cerebro sea un arcano, el universo, reflejo de su estructura, será un arcano también.

(As long as our brain is a mystery, the universe, the reflection of the structure of the brain, will also be a mystery.)

—SANTIAGO RAMÓN Y CAJAL (1922)[1]

The words of pioneering neuroscientist Santiago Ramón y Cajal (1852–1934) reflect the thoughts of many scientists and philosophers who over the centuries have tried to unravel the workings of the brain. Using the most advanced tools and technologies available in their time, these men and women have figuratively and literally probed the brain, seeking treatments and cures for neurological disease, and searched for the underlying cause of what makes us human.

Scientists take many different pathways to the field of brain research and come from diverse academic backgrounds. A neuroscientist may enter the field with a medical degree or a doctoral degree in neuroscience, biology, chemistry, physics, bioengineering, physiology, or a related field. Others approach neuroscience from outside the natural or physical sciences, for example, graduating from college with a major in music or philosophy. For me, neuroscience was not on my list of career aspirations until my last years of college. As a kid, I grew up in Los Angeles (California), Kuala Lumpur (Malaysia), and Kobe (Japan). Whether it was riding my bike along the concrete and asphalt streets of Southern California, searching the monsoon drains surrounding my house in Southeast Asia, or climbing the hills behind my school in Kobe, I tried to get outside as much as possible. Perhaps my love of the outdoors is why my high school career aptitude test determined that I was best suited to become a forester.

Both of my parents were teachers and, although neither of them had any background in science, they always encouraged me to follow my interests regardless of the subject. In high school, my interests were mainly sports. But during my senior year of high school, I took a marine biology class where students were taught about the invertebrates, birds, fish, and mammals that live in and around the ocean. This class sparked my curiosity about nature and motivated me to learn more about marine life. On weekends, I would drive thirty minutes to the beach in Los Angeles—not to sit on the sand, but to explore the tide pools. Each day there revealed something new and unexpected. I spent hours turning over rocks (and carefully putting them back in place) to see the invertebrate life that lived hidden at the intersection of sand and sea. These days I check the tide tables and get out to the tide pools of Puget Sound several times a year.

When I entered college at the University of California, Los Angeles (UCLA), in 1976, I thought that I would become an oceanographer. My favorite course, invertebrate biology, included trips to the bays around Los Angeles, where students would board a boat and dredge the ocean floor for moon snails, sea cucumbers, worms, and other animals. We would take our catch back to the lab to study our specimens. During class, the professor entertained students with stories of how he spent his research time studying shrimp in warm South Pacific Ocean waters. This sounded like a great career: to get paid for something I would do for free.

My academic pathway took a new direction at the start of my junior year at UCLA when I enrolled in an introductory psychology class. The instructor of the class was Dr. John Liebeskind (1935–1997), who introduced himself as a physiological psychologist but these days would be called a neuroscientist. I didn't know it at the time, but Dr. Liebeskind was an innovative researcher who was making important discoveries about the brain's own pain inhibitory system. After one lecture about the brain, Dr. Liebeskind invited students to follow him back to his lab for a short tour. I joined a small group of three or four students who took him up on his offer and followed the professor to his lab in the basement of the psychology building. At the end of the brief tour, Dr. Liebeskind said that any of us students could join his lab if we showed up the following day. I was the only student who returned the next day and was put to work quickly by the graduate students and postdoctoral researchers. There was no undergraduate neuroscience major at UCLA until 1992, so instead I switched my major to psychobiology. I worked in the Liebeskind lab as a volunteer undergraduate researcher until I graduated in 1980 with my bachelor's degree in psychobiology. I then went on to pursue a master's degree and a doctorate degree before taking a research position at the National Institutes of Health in Bethesda, Maryland, where I studied how the brain processes information related to touch and pain. I ultimately landed at the University of Washington, where I've been since 1991.

Neuroscience is important not only for researchers in labs and physicians in clinics, but for everyone. We are all likely to know someone affected by a neurological disorder. Alzheimer's disease, Parkinson's disease, multiple sclerosis, depression, autism, stroke, schizophrenia—the list of neurological and psychiatric disorders goes on and on. These diseases take a tremendous emotional and economic toll on patients, family members, and caregivers. If we all knew more about the brain, perhaps we could better empathize with people affected by these disorders, reduce stigma attached to neurological diseases, and help people cope with their conditions. Beyond a better understanding of disease and disorders, the rapid advances in neuroscience make it imperative that everyone has a basic understanding of brain research to read articles in magazines, newspapers, and websites. It is difficult nowadays to read through popular media without seeing a story discussing the brain. Unfortunately, there is also a considerable amount of misinformation and misunderstanding about the brain. For example, a commonly held belief is that we use only a small portion of the brain; we actually use all of the brain. People with an understanding of brain research should be better equipped to analyze information in the media more critically. Further, new neuroscientific discoveries are impacting many segments of society, including our courtrooms, classrooms, and companies. Lawyers are using brain imaging to sway juries, teachers are looking to neuroscience for help to improve educational practices, and industry is developing new methods to enhance brain function. A neuroscientific-literate society will be better prepared to debate how such advances should be developed and disseminated.

In this book, I hope readers will enjoy learning about concepts, terms, structures, and people associated with the field of neuroscience. A complete catalog and detailed description of every neuroscientific phrase, brain structure, and influential neuroscientist is not possible in these pages, but I have tried to include entries that will stimulate your curiosity, help you become more familiar with neuroscience, and motivate you to learn more about the brain. Knowing neuroscientific facts and figures is only a small part of learning about the brain, and neuroscientists do not claim to fully understand how the brain works. Rather than just listing facts and figures, this book should provide you with a basic understanding of how neuroscientists have built theories about the brain and how the field has progressed over time.

Neuroscientific research has provided new treatments and even cures for some brain diseases, but effective therapies for many neurological diseases remain elusive, and the underlying mechanisms of some basic neurological functions (e.g., consciousness) are still not known.

Don't get me wrong. Researchers have made monumental progress in our understanding of the nervous system in health and disease and have provided exquisite detail about the anatomy and physiology of the nervous system, in thousands of papers published every year. These publications provide the pieces of the puzzle to a better understanding of ourselves and our place in nature—but many pieces are still missing.

I hope this book will answer some questions you have about the brain and nervous system. My intention is to open a window to the process of discovery that occurs in research laboratories every day. Answers to the mysteries about the brain are core to who we are as human beings.

Students often ask me what I like about being a neuroscientist. My response to this question is the same as why I turn over rocks in tide pools: you never know what you will find. Although we know an incredible amount about how the brain works, there is still so much more to learn. I hope *Neuropedia* will motivate you to turn over some rocks.

Neuropedia

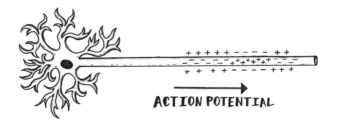

ACTION POTENTIAL

Action Potential

Electrical signal that is the basic unit of communication for transmitting information throughout the nervous system. Capable of traveling at speeds faster than 250 mph, action potentials whisk electrochemical messages from neurons to muscles, organs, and other neurons to control our movements, emotions, perceptions, thoughts, and actions.

Particles in our body that are electrically charged are called ions. To set up an action potential, positively charged sodium ions and potassium ions and negatively charged chloride ions and protein molecules are arranged so there are different amounts of these ions on different sides of a semipermeable neuron membrane.

This creates an electrical potential difference between the inside and outside of a neuron. When a neuron is not sending a signal, the inside of a neuron is negative relative to the outside because sodium ions do not move easily across the membrane and a protein pump sends three sodium ions out of the neuron for every two potassium ions it puts in. In fact, the inside of most neurons is about 70 mV more negative relative to its outside.

When a neuron sends a signal down an axon, the rapid exchange of sodium and potassium ions across a neuron's membrane creates an electrical signal called an action potential. During an action potential, sodium ions rush into a neuron through channels (openings) in the neuron membrane. This causes the inside of a neuron to become more positive. A short time later, different channels open to let potassium ions flow out of a neuron across the neuronal membrane, and the sodium channels start to close. This causes the neuron to return to the more negative state. Eventually, the concentrations of all ions go back to their original levels inside and outside of the neuron. The entire process takes only a few milliseconds. Also, the action potential is "all or none"—meaning that if it gets started, it will remain at the same size as it moves down the axon.

Alan Lloyd Hodgkin (1914–1998) and Andrew Fielding Huxley (1917–2012) shared the 1963 Nobel Prize in Physiology or Medicine for their foundational research about the action potential. These scientists used the giant axon of the squid to learn how ions flow across the neuronal membrane to create an action potential.

See also Axon, Squid Giant; Neuron; Neurotransmitters

Ageusia

Complete loss of the sense of taste. Although the inability to taste is not fatal, it can cause a lower appetite and the loss of one of life's simple pleasures. Many people have a reduction in their sense of taste as they get older, but it is rare that someone loses their ability to taste completely.

Ageusia can be caused by damage, infection, or injury to the sensory nerves of the tongue (facial nerve and glossopharyngeal nerve). Radiation therapy used to treat cancer in the head or neck can result in ageusia by damaging taste buds, nerves, or salivary glands. Ageusia is also a known side effect of some antibiotic, stimulant, and antipsychotic medications. Some people regain their sense of taste when they stop taking a particular medication or when an injury heals.

The loss of taste is included on the symptom list for a severe acute respiratory syndrome coronavirus-2 (SARS-CoV-2) infection (COVID-19) by public health organizations. Although the most commonly reported symptoms of COVID-19 infections are fever, cough, and fatigue, severe reductions in taste have been reported in about 40% of COVID-19 patients.[2]

See also Coronavirus Disease 2019 (COVID-19); Cranial Nerves

Alcohol

Central nervous system depressant created by fermenting grain, fruit juice, or honey. Alcohol (ethanol) has been consumed by people for thousands of years, making it the world's oldest drug. After alcohol is consumed, it enters the bloodstream through the stomach and small intestine. The heart then pumps the alcohol to the brain, where it acts to produce relaxation, lower inhibitions, slow reflexes and reaction time, and affect coordination. In other words, it makes people drunk. Although small amounts of alcohol may cause people to make fools of themselves, excessive consumption of alcohol can result in trouble breathing, loss of consciousness, and even death.

The alcohol molecule is small and soluble in lipids and water, so it crosses the blood-brain barrier easily. Chronic alcohol use can reduce brain size, increase cerebral ventricle size, and cause a B1 (thiamine) vitamin deficiency. A thiamine deficiency may cause Wernicke's encephalopathy or Korsakoff's syndrome—neurological disorders characterized by memory impairment, confusion, and movement disorders. Infants exposed to alcohol during development may be born with fetal alcohol syndrome.

Alcohol affects multiple brain areas and several neurotransmitter systems. Alcohol binds to the receptors for serotonin, glutamate, and GABA and leads to a release of dopamine. The symptoms of intoxication are the result of alcohol's effect on a specific area of the brain. For example, alcohol interferes with the ability of the (1) hippocampus to form memories, (2) cerebellum to control balance, and (3) frontal lobe to manage judgment. The overall effect of alcohol is to depress the nervous system.

Humans are not the only animals with a taste for alcohol. Pen-tailed tree shrews drink alcohol from fermented bertam palm flowers.[3] However, these small mammals from Southeast Asia show no signs of drunken behavior and do not get into bar fights.

See also Blood-Brain Barrier; Cerebellum; Dopamine; Fetal Alcohol Syndrome; Frontal Lobe; GABA; Hippocampus; Serotonin

Alien Hand Syndrome

Rare neurological disorder characterized by involuntary movement of a hand accompanied by the perception that the movement is not controlled by its owner. In 1964, Stanley Kubrick (1928–1999) produced and directed the film *Dr. Strangelove* with scenes of its title character, played by Peter Sellers (1925–1980), losing control of his right arm while his left arm tries to restrain it. This film gave alien hand syndrome the alternative name Dr. Strangelove syndrome.

People with alien hand syndrome make purposeful movements of their hands, but they believe that they are

not controlling their limbs and act as if the limb does not belong to them.[4] Some people have even given names to the "alien" hands. The condition may arise after stroke, trauma, or disease of the corpus callosum, frontal lobe, or parietal lobe. One theory proposed for the cause of alien hand syndrome is that brain regions responsible for planning and controlling movement are disconnected. When these connections are lost, these different brain areas work independently so that physical movement of a body part is no longer associated with the conscious perception of actually controlling the body part.

Most treatments for alien hand syndrome focus on behavioral therapy to help people cope with their condition. Such therapy can provide ways for people to regain control of their daily lives. Some people with alien hand syndrome benefit from learning how to distract themselves from their affected hand or to visualize taking control of their hand.

The first mention of alien hand syndrome was made by Kurt Goldstein (1878–1965) in 1908. Goldstein described the case of a fifty-seven-year-old woman whose left hand grabbed her own neck and choked her.

See also Corpus Callosum; Frontal Lobe; Parietal Lobe

Alzheimer's Disease (AD)

A progressive, degenerative brain disease characterized by memory loss and disorientation. When is forgetfulness a normal sign of aging and when is it a symptom of a neurological disorder, such as AD? Everyone sometimes forgets where they placed their keys or cannot remember an item on a shopping list. Those types of memory loss do not affect a person's daily life. However, memory loss that disrupts a person's ability to work, communicate properly, control emotions, and make decisions may be a sign of AD. Dementia, the characteristic feature of AD, is the general term for this gradual and persistent loss of cognitive, linguistic, and emotional abilities.

The vast majority of people who develop AD are over the age of sixty-five years. Although the specific causes of AD have not yet been identified, genetic and environmental factors are likely involved. AD results in a slow

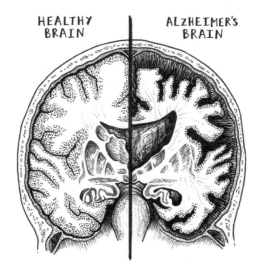

HEALTHY BRAIN | ALZHEIMER'S BRAIN

and irreversible attack on the brain that destroys neurons and the connections between neurons. This brain damage appears to be caused by abnormal buildup of beta-amyloid proteins called plaques and tau proteins called tangles. The accumulation of plaques and tangles may trigger inflammation, interfere with messaging between neurons, and kill neurons. The cerebral cortex and hippocampus, areas of the brain involved with thinking, learning, language and memory, are especially damaged.

Although there is no cure for AD, some medications and therapies can treat symptoms and improve the quality of life of people who suffer from AD. For example, some drugs that target the acetylcholine or glutamate neurotransmitter systems can help with memory loss and other signs of dementia. Nonpharmacological treatments, such as behavioral therapy, can help some people with AD gain control of their daily activities.

Celebrities who have been diagnosed with AD include singers Glenn Campbell (1936–2017), Perry Como (1912–2001), and Tony Bennett (1926–); basketball coach Pat Summitt (1952–2016); actors Charles Bronson (1921–2003), James Doohan (1920–2005), Charlton Heston (1923–2008), Rita Hayworth (1918–1987), Burgess Meredith (1907–1997), Estelle Getty (1923–1998), Peter Falk (1927–2011), James Stewart (1908–1997), and Eddie Albert (1906–2005); former US president Ronald Reagan (1911–2004); painter Norman Rockwell (1894–1978); and boxer Sugar Ray Robinson (1921–1989).

See also Cerebral Cortex; Hippocampus; Neurotransmitters

Amphetamine

A synthetic central nervous system stimulant. Amphetamines are a group of stimulant drugs that include Benzedrine, dextroamphetamine, and methamphetamine. The first amphetamines were developed to treat asthma, sleep disorders (narcolepsy), and hyperactivity. These drugs were used during World War II and later military conflicts to keep soldiers and pilots alert and to fight off combat fatigue.[5]

Amphetamines excite the central nervous system and sympathetic division of the peripheral nervous system primarily by increasing the activity within the dopamine and norepinephrine neurotransmitter systems. For example, amphetamines (1) cause the release of dopamine and norepinephrine from axon terminals, (2) block dopamine and norepinephrine reuptake, (3) inhibit the storage of dopamine in synaptic vesicles, and (4) inhibit the destruction of dopamine by enzymes. These actions increase the availability of dopamine and norepinephrine within synapses where the neurotransmitters can bind to receptors.

People who use amphetamines usually experience a rise in heart rate and blood pressure, reduced appetite, widening of their pupils, and increase in alertness soon after the drug is ingested, smoked, or injected. Prolonged use of amphetamines can result in sleep disturbances, hallucinations, and tremors. Some people also become addicted to the effects of amphetamines and show tolerance to the drug when they must increase the amount they take in order to feel the effects.

The US Drug Enforcement Agency currently classifies amphetamines as Schedule II stimulants, meaning that products approved by the Food and Drug Administration that contain amphetamines have acceptable medical uses, but they also have a high potential for abuse.[6]

See also Autonomic Nervous System; Narcolepsy; Neurotransmitters; Synapse

Amygdala

Almond-shaped brain structure located in the temporal lobe of the brain. As part of the limbic system, the amygdala has important roles in emotional behavior, memory, anxiety, and fear. Neurons in the amygdala respond to emotional faces and unpleasant odors, tastes, and feelings.[7] These complex functions require that neurons in the amygdala process information from all the senses and from organs inside the body through input from the hypothalamus, thalamus, cerebral cortex, and brainstem areas. Signals sent from the amygdala feed back to the hippocampus, thalamus, and brainstem.

The role of the amygdala in fear was discovered when researchers observed that animals with damage to the amygdala did not learn to be afraid of new fears. This observation suggests that the amygdala helps organisms learn and remember the emotional significance of an event. Fear is not the only emotion processed by the amygdala. Some portions of the amygdala are involved with motivation, reward, aggression, maternal behavior, and sexual behavior.

The importance of the amygdala and its relationship to emotional behavior is strengthened by the behavior of people with a strange neurological disorder called Klüver-Bucy syndrome. Klüver-Bucy syndrome develops after people have suffered damage to their temporal lobes, including the amygdala and hippocampus, on both sides of their brain. People with Klüver-Bucy syndrome display hyperorality (putting things in their mouth), hyypermetamorphosis (touching everything they see), hypersexuality, amnesia, and lack of fear and anger. There is no cure for Klüver-Bucy syndrome, but some people can learn to manage their symptoms.

See also Hippocampus; Temporal Lobe

Amyotrophic Lateral Sclerosis (Lou Gehrig's Disease)

Progressive and fatal neurological disease characterized by a slow degeneration of brain and spinal cord neurons that control movement. In 1939, baseball star Lou Gehrig (1903–1941) knew that his performance on the field was suffering. He did not have his usual hitting power and he had difficulty running the bases.

Later that year, the problem was identified: Gehrig had amyotrophic lateral sclerosis (ALS). Even today, ALS is best known as Lou Gehrig's disease, named after the longtime New York Yankee first baseman.

When motor neurons die, messages are no longer sent from the brain and spinal cord to control muscles for movement. The death of motor neurons causes progressive paralysis, leaving a person unable to move. Eventually, a person with ALS has trouble speaking, breathing, and eating. ALS usually does not affect memory, personality, or the senses. The disease is not contagious and there is no known cause for the majority of cases, but an inherited form of ALS is responsible for 5%–10% of the cases. Unfortunately, there is no cure for ALS, but some drugs (e.g., riluzole, edaravone) and therapies can improve the quality of life for people suffering from the disease.

In addition to Gehrig, physicist Stephen Hawking (1942–2018), actor David Niven (1919–1983), actor/playwright Sam Shepard (1943–2017), football player Dwight Clark (1957–2018), and former senator Jacob Javits (1904–1986) were all diagnosed with ALS.

Aplysia (Sea Hare)
Marine mollusk used by neuroscientists to study neuron function. The sea hare (*Aplysia califonica*) deserves a special honor in the archives of neuroscience for its contributions to our understanding of the nervous system. With approximately ten thousand neurons in its entire body (compared to eighty-six billion neurons in the human brain), the *Aplysia* has provided neuroscientists with a model organism to study the neural basis of behavior, especially memory and learning.

Starting his work in the 1960s, neuroscientist Eric Kandel used *Aplysia* to study the neuronal mechanisms responsible for the siphon withdrawal reflex.[8] This response involves the withdrawal of the animal's siphon (a tube that directs water out of the body) when it is touched. The neuronal circuit responsible for the withdrawal response is relatively simple, and the neurons in the pathway are large and easy to find in different *Aplysia*. The behavior can also be modified by learning. Using this system, Kandel and his coworkers provided an excellent way to demonstrate how learning and memory affect the strength of synaptic connections between neurons.

Kandel was rewarded for his work about neuronal signal conduction with the 2000 Nobel Prize in Physiology or Medicine.

See also Synapse

Aristotle (384–322 BC)

Greek philosopher and student of Plato who expounded on topics including biology, physics, logic, politics, and the arts. Aristotle's teachings set the foundation for Western philosophy and the method of scientific inquiry.[9]

Although Aristotle is without question a towering historical figure who made important contributions to Western culture and science, he was wrong about the basic functions of the brain. Aristotle believed that the brain served merely to cool heat generated by the heart. He argued that the heart, not the brain, was responsible for intelligence, consciousness, and sensation. Aristotle's observations that the heart was located in the middle of the body and that it was the first organ to develop in chicken embryos may have contributed to his cardiocentric bias. The great philosopher also made errors about neuroanatomy. For example, he stated that the brain does not fill the entire cranium and that the back of the head was empty. These mistakes may be chalked up to the likelihood that Aristotle never dissected a human body to look inside. He may have based his neuroanatomical knowledge on studies of nonhuman animals, such as fish and reptiles.

Although we no longer think of the brain simply as a refrigerator or radiator to cool blood and the heart as the seat of intelligence and emotion, we nevertheless pay respect to this long-ago way of thinking each time we learn "by heart" or send our "heartfelt thanks" or suffer a "broken heart."

Attention Deficit Hyperactivity Disorder (ADHD)

Common neurodevelopmental condition characterized by inattention, hyperactivity, and impulsivity. ADHD affects millions of children each year, and the symptoms of the disorder can continue into adulthood.

Behavior that may indicate that a child has ADHD includes inattention (being easily distracted; being forgetful; problems following directions), hyperactivity (problems sitting still), and impulsivity (inability to control impulses; acting without thinking). Because children may show these behaviors for reasons other than ADHD, a clinician must examine a child carefully before making a diagnosis.

The exact cause of ADHD is not known, but research suggests that genetic and environmental factors likely contribute to the condition. The possibility of a genetic link is strengthened by research showing the incidence of ADHD is more common in identical twins than in fraternal twins. Other factors that increase the chances that a child will develop ADHD include low birth weight, prenatal exposure to toxins, and brain injury.

Medications and behavioral therapy are often used to treat the symptoms of ADHD, but there is no cure. Surprisingly, central nervous system stimulants (e.g., Ritalin) can often reduce hyperactivity and impulsivity.[10] The beneficial effects of these drugs may be related to their ability to increase brain levels of the neurotransmitter dopamine, which is important for attention. Therapy to teach children and adults ways to keep organized and manage behavior can also help people function in school and at work.

Intelligence is not linked to ADHD, and the condition is not caused by watching too much TV, food allergies, or eating too much sugar.

See also Dopamine

Autism Spectrum Disorder (ASD)

Developmental brain condition affecting how people socialize and communicate with other people. The US

Centers for Disease Control and Prevention estimate that one out of every fifty-four children is born with autism spectrum disorder.[11] The severity of symptoms of ASD varies from person to person, but the characteristic signs of the condition include communication problems, repetitive movements, and difficulty with social interaction.

People with ASD may experience difficulties in verbal communication, and some are entirely nonverbal; they may also have trouble interpreting what other people are saying or doing. Health care professionals diagnose ASD by looking for these behaviors and ruling out other disorders.

The exact cause of ASD is not known, but multiple factors appear to be responsible. Strong evidence supports a genetic factor as a contributor to ASD. Identical twins are more likely to both have ASD than either fraternal twins or nontwin siblings. A genetic variable may make someone more susceptible to environmental factors, such as a chemical or infection that may cause ASD.

Treatments for ASD seek to reduce the symptoms of the disorder and help people with their daily activities. Therapies to manage behavior sometimes help to improve skills and reduce unwanted behaviors. Cognitive therapy can help people with autism identify thoughts and feelings that lead to problem situations. Although there is no drug to cure ASD, some antidepressant, antipsychotic, stimulant, antianxiety, and anticonvulsant medications can reduce some symptoms of ASD.

Rain Man, a 1988 film starring Dustin Hoffman and Tom Cruise, tells the story of a man who has ASD in combination with savant syndrome (exceptional memory, rapid calculating ability). The film raised awareness about ASD in popular culture, but received some criticism about the depiction of savant syndrome because only about one in ten people with ASD show those extraordinary abilities.[12]

Autonomic Nervous System

The part of the peripheral nervous system that helps maintain functions of internal organs, such as digestion, respiration, and heart rate. The autonomic nervous system is composed of the sympathetic, parasympathetic, and enteric nervous systems. The autonomic system works in the background because it functions in an involuntary and reflexive manner; it comes into play during emergencies ("fight or flight" situations) and in nonemergencies ("rest and digest" situations).

When a person walks around a corner and comes face-to-face with a snarling dog, should the person run away or prepare to fight? The heart pounds faster, blood pressure rises, digestion of food slows. That is the sympathetic nervous system in action. The sympathetic nervous system is activated when the body mobilizes for defense or in response to stress. In defensive situations, the heart rate increases, the lungs expand to hold more oxygen, the pupils dilate, and blood flows to the muscles.

Lounging at the park, resting on a bench, enjoying the sunshine: in these situations, a person's digestion kicks into gear, blood pressure decreases, and the heart rate slows. That's the parasympathetic nervous taking control. The enteric system is a network of neurons in the gut that also helps regulate digestion.

In general, the functions of the sympathetic nervous system and parasympathetic nervous system work in opposition to each other. However, these systems are always working to keep the body in balance.

Avicenna (980–1037)

Persian physician. Avicenna was a medical doctor who contributed to our understanding of the anatomy and physiology of the nervous system as well as the diagnosis and treatment of neurological and psychiatric disorders. Avicenna's primary work, *The Canon of Medicine*, includes the diagnoses and treatments of disorders such as epilepsy, stroke, headache, meningitis, and head injury.[13] *The Canon of Medicine* was used as a medical guide for hundreds of years after Avicenna's death.

Avicenna may have been centuries ahead of his time in the approach he took to treat disease. He recommended a change in diet and advocated for physical exercise as therapy for many diseases. He also emphasized the importance of sleep in managing psychiatric disorders. In addition to suggesting changes in lifestyle to manage disease, Avicenna prescribed herbs or other medications for neurological conditions. Perhaps the first use of electrical stimulation to treat psychiatric disorders can be attributed to Avicenna: he suggested that depression could be cured by placing a live electric fish (torpedo ray) onto a patient's forehead.

Axon

The part of the neuron that transmits information away from the cell body. The axon, which is connected to the cell body by the axon hillock, sends action potentials toward synaptic terminals. Some axons are very short (less than 1 mm), while others, such as one stretching from the spinal cord to the foot, can be very long (~1 m; 3.3 ft). The diameter of an axon also varies in size from about 0.1 to 20 microns.

Larger-diameter axons conduct action potentials at faster speeds than smaller-diameter axons. To increase the transmission speed of action potentials, some axons are surrounded by myelin insulation. Action potentials in small-diameter, unmyelinated axons travel at speeds between 0.5 and 2.0 m/s (1.8–7.2 km/hr) while action potentials in large-diameter, myelinated axons travel at speeds between 80 and 120 m/s (288–432 km/hr).

See also Action Potential; Axon, Squid Giant; Myelin; Saltatory Conduction; Synapse

Axon, Squid Giant

Large extension from a squid's nerve cell body. Like the sea hare (*Aplysia*), the squid is another invertebrate that deserves our gratitude for its important contributions to our understanding of the nervous system. The giant axon of the squid, part of the animal's water jet propulsion system, has been used to help neuroscientists understand how neurons send electrical signals.[14] These giant axons are so large (0.3–1.0 mm in diameter) that they can be seen without a microscope. The large size and location of giant axons make it easy for scientists to study these nerve fibers.

In the 1930s, English physiologist John Zachary (J. Z.) Young (1907–1997) described the structure of the giant axon in the squid *Loligo*. Later, Alan Lloyd Hodgkin (1914–1998) and Andrew Fielding Huxley (1917–2012)

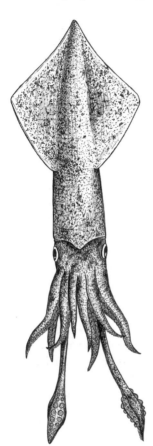

used the squid giant axon in their 1963 Nobel Prize–winning work to describe how action potentials are conducted. Because of the large size of the squid giant axon, an electrode can be placed completely inside the giant axon to measure the voltage difference between the inside and the outside of the axon. Also, the cytoplasm of the axon can be squeezed out like toothpaste from a tube and replaced with solutions containing different concentrations of ions (e.g., sodium, potassium, chloride). Using this experimental setup, scientists were able to demonstrate how the exchange of ions across a neuron's membrane was responsible for the conduction of action potentials.

Later experiments with toxins and drugs such as tetrodotoxin (TTX) and tetraethylammonium (TEA), which block the exchange of sodium and potassium ions, respectively, confirmed the importance of these ions and set the foundation for our understanding of the generation and propagation of action potentials.

So the next time you are enjoying a plate of calamari, give thanks to the squid for its contributions to neuroscience.

See also Action Potential; *Aplysia*; Neurotoxin

Blind Spot

Area of the retina lacking photoreceptors; also called the optic disk or optic nerve head. The retina of the eye contains cells (photoreceptors) that respond when they are exposed to light. The photoreceptors are connected to other cells that send their axons in the optic nerve to the brain. However, one small place on the retina is devoid of photoreceptors because axons of the optic nerve and blood vessels that exit the eye take up space. Light that falls on this area does not strike any photoreceptors, and therefore there are no signals sent to the brain about that light.

With two eyes, having a blind spot is not an issue because light strikes different areas of each retina so the brain receives a complete picture. Even with one eye, a blind spot rarely causes any problems because the brain fills the information gap with what it expects to be there.

How the brain makes up for missing information can be demonstrated simply by centering objects on the blind spot (see illustration). To find your blind spot, hold this book about 50 cm (20 in) from your face, close your right eye and look at the + sign in the image with your left eye. You should be able to see the O on the left side in your peripheral vision. Slowly move the

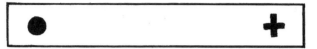

book closer to your face while you focus on the +. At some point the O will disappear. When it does, you have focused the image of the black circle on your blind spot.

See also Retina

Blood-Brain Barrier (BBB)

A semipermeable system of astrocytes and capillaries in the brain that restricts the flow of some substances from crossing out of the bloodstream into the brain's circulatory system.

Think of the BBB as the brain's border guard, allowing entry to some materials and restricting entry to others. Endothelial cells that line capillaries in the brain fit together tightly to reduce the flow of large molecules, low fat-soluble molecules, and high electrically charged molecules across blood vessels. Glial cells (astrocytes) may help in the development and maintenance of the BBB. The restrictive barrier provided by this system protects the brain from substances in the blood that may damage the brain. The BBB also regulates brain levels of hormones and neurotransmitters that are released into the blood from other parts of the body and maintains a stable chemical environment for the brain.

The BBB can be weakened (opened) by hypertension, high concentration of a substance in the blood, microwaves, radiation, infection, trauma, ischemia, and inflammation. Also, before birth, the BBB is not fully developed. Several areas of the brain, called circumventricular organs, have a BBB where substances can more easily cross into the brain. Circumventricular organs include the pineal body, posterior pituitary, area postrema, subfornical organ, vascular organ of the lamina terminalis, and median eminence.

See also Glia

Brain, Development

Prenatal and postnatal growth and modification of the brain. The brain originates from embryonic tissue called the ectoderm. Approximately two weeks after conception, the ectoderm forms a neural plate. Within a week or so, a fold is formed in the neural plate to create a neural groove. By three weeks, the edges of the neural groove fold to form the neural tube. The front of the neural tube goes on to develop into the brain, and the remainder of the neural tube develops into the spinal cord.

A newborn baby's brain weighs just under 400 gm (0.9 lb) and contains almost all of the neurons it will ever have. In fact, adults have fewer neurons than babies. During development, neurons are overproduced. As a child grows, those neurons that are not used die. Although some neurons in a few parts of the brain, such as the hippocampus, may develop after childhood, most neurons are not replaced when they die. The brain continues to grow after birth with the addition of glial cells that divide and multiply. The average adult human brain weighs about 1,400 gm (3 lb).

During some prenatal stages of development, the brain adds about 250,000 neurons every minute. By the age of two years, a child's brain is about 80% of its adult size.

Brain Initiatives
International efforts to better understand the physiology and anatomy of the brain, develop new methods to study the brain, and discover new therapies and treatments for neurological disorders. On April 2, 2013, then president Barack Obama stood at a lectern in the White House and announced the start of the Brain Research through Advancing Innovative Neurotechnologies (BRAIN) Initiative. With an initial federal investment of $110 million and additional financial contributions from several private foundations, BRAIN Initiative researchers would develop new methods to investigate how neurons interact to understand how brain function is linked to behavior. A few months after the BRAIN Initiative began, the Human Brain Project (HBP) started with funding from the European Union. Like the BRAIN Initiative, the HBP was established to create new technologies to better understand brain function and its relationship to cognition. Soon the governments of Australia, Canada, China, Japan, and Korea followed with their own efforts to support brain research and discover new ways to treat neurological disease. These large-scale brain research programs not only invested resources into basic scientific studies, but also were concerned with ethical and societal issues that may arise from their discoveries.

The various brain initiatives have created important innovations, such as new technologies to stimulate and record from the nervous system, new databases of neuron cell types, novel brain maps (brain atlases), and an ethical framework to design and conduct neuroscientific research. However, the tremendous progress made through these efforts has not been without controversy.[15-18] For example, the share of project funding available to individual researchers compared to large collaborative funding has come under scrutiny. The European Union HBP has received criticism for setting unrealistic goals, poor organization and leadership, and wasteful spending.

Although it is difficult to measure the overall impact of these large research projects, these brain initiatives do provide hope for the millions of people affected by neurological disorders.

See also Neuroethics

Brainstem

The central core of the brain, connecting the spinal cord and brain. The brainstem is composed of the medulla, pons, and midbrain. Together these areas help regulate basic life processes, such as respiration, heart rate, sleep cycle, and digestion. Damage to the brainstem, for example, by trauma or stroke, is often lethal. The brainstem also contains the pathways that send information between the brain and spinal cord.

The top of the spinal cord gradually merges with the medulla (medulla oblongata). The top of the medulla forms the bottom of the fourth ventricle, a chamber containing cerebrospinal fluid. Areas within the medulla are responsible for regulating respiration, heart rate, and blood pressure and controlling reflexes for swallowing, vomiting, coughing, and sneezing. The pons is located just in front of the medulla. Some neurons in the pons function to control sleep while other neurons process sensory information from the head or send signals to control muscles for eye movement, facial expression, chewing, and swallowing. The midbrain is found above the pons. Within the midbrain are two paired structures, called the superior colliculus and the inferior colliculus, that are important for processing visual and auditory information, respectively. The substantia nigra, a major source that produces the neurotransmitter dopamine, and the raphe nuclei, a major source of the neurotransmitter serotonin, are also found in the midbrain.

The pons gets its name from the Latin word meaning "bridge" because of its connection between the medulla and the midbrain. The medulla also comes from a Latin word meaning "marrow" because of the inner location of the structure. Although the names of areas within the midbrain were derived from Latin (for example, colliculus means "small hill"), the midbrain just means "middle brain."

See also Dopamine; Serotonin

Broca, Paul (1824–1880)

French neurologist. In the mid-1800s, debate raged about whether the cerebral hemispheres functioned as a single unit or if different areas were specialized for different functions. Broca was fortunate to study a patient who contributed to the debate.

In 1861, a man named Louis Victor Leborgne (1809–1861) was admitted to the hospital where he was cared for by Broca.[19] As a youth, Leborgne suffered from epilepsy and he lost his ability to speak when he was thirty years old. Although Leborgne could understand spoken language, he could speak just one word: tan. For this reason, many textbooks refer to Leborgne as Tan.

When Leborgne was fifty years old, he was paralyzed on the right side of his body, developed gangrene, and was transferred to Broca for care. Leborgne died just six days later. When Broca removed and examined Leborgne's brain during an autopsy, he found damage to the left cerebral hemisphere in the frontal lobe. In 1861, Broca made a presentation to the Society of Anthropology in Paris, where he proposed that the area of the brain damaged in Leborgne was responsible for the creation of speech. Broca continued to study other patients who could not speak and confirmed that these patients also suffered damage to the left cerebral hemisphere.

Broca's observations provided evidence that a specific region of the brain had a specific function. The area of the brain damaged in Leborgne and the other patients is now known as Broca's area, and the difficulty in producing speech is called Broca's aphasia. We now know that, to speak a word, information must first get to an area of the cortex (visual cortex for speaking a written word; auditory cortex for speaking a heard word) and then be transmitted to Wernicke's area. From Wernicke's area, information travels to Broca's area, then to the motor cortex.

See also Frontal Lobe; Wernicke, Carl

Caffeine

Central nervous system stimulant in the xanthine chemical group; found naturally in coffee beans and tea leaves and added to some soft drinks and drugs. After caffeine is consumed, it is absorbed in the stomach and small intestine and then transported in the blood to the brain. In various areas of the brain,

caffeine interferes with the action of the neurotransmitter adenosine. Modification of adenosine's action can result in increased alertness and attentiveness. In other parts of the body, caffeine can increase heart rate, constrict blood vessels, and improve breathing.

Depending on the type of coffee and the brewing method, a cup of coffee has between 60 and 150 mg of caffeine. The side effects of caffeine, such as insomnia, headaches, and nervousness, are well known to many people who enjoy their regular mug of coffee. Similarly, some people experience uncomfortable withdrawal symptoms when they suddenly stop their usual consumption of caffeinated products. Some people develop a tolerance to caffeine and must consume greater amounts of caffeinated beverages to achieve the same effects. Genetic factors appear to partially contribute to the tolerance to caffeine.[20] Massive doses of caffeine (about 10 grams or 80–100 cups of coffee) can be deadly.

The discovery of coffee is shrouded in mystery, but one legend claims that goats were responsible for unearthing the power of this stimulant. According to the story, around AD 850, goats belonging to an Egyptian goat herder named Khaldi did not return home one night. When Khaldi eventually found his goats, they were dancing around a shrub with red coffee beans. Like any conscientious goat herder, Khaldi tried some of the berries and he started to dance. After a bit of experimenting with brewing berries, coffee was born.

See also Neurotransmitters

Capgras Syndrome

Rare neurological condition in which people believe that family members and friends are impostors; also called impostor syndrome.[21] Imagine that your spouse or significant other insists that you are not who you think you are, instead maintaining that you are an impostor who has replaced the "real" you. Such is the life of a person who lives or takes care of someone with Capgras syndrome.

Although Capgras syndrome is somewhat rare, people with dementia, Alzheimer's disease, Lewy body disease, Parkinson's disease, epilepsy, stroke, and schizophrenia make up the majority of cases with the disorder.[22, 23] A disconnection between brain areas involved with facial recognition and emotions is theorized to underlie the problem.

Capgras syndrome is not like prosopagnosia (face blindness). People with Capgras syndrome recognize a face, but they have the delusion that the face belongs to an impostor or double. Pets and inanimate objects are sometimes included in a delusion. Reasoning with people with Capgras syndrome does not eliminate the misbelief.

Antipsychotic or antidepressant medications or drugs used to treat dementia may reduce symptoms of Capgras syndrome in some people. Behavioral therapy can help patients feel at ease with people they see as impostors. Family therapy and counseling can also help a patient's loved ones cope with the condition.

French psychiatrist Joseph Capgras (1875–1950) and Jean Reboul-Lachaux first described the disorder in 1923.

See also Alzheimer's disease; Epilepsy; Lewy body disease; Parkinson's disease; Prosopagnosia; Schizophrenia; Stroke

Cauda Equina

Bundle of spinal nerves at the base of the spinal cord. From the Latin words for "horse's tail," the cauda equina is a collection of nerves extending from the spinal cord that sends and receives information from skin and muscles of the lower limbs, bladder, rectum, and reproductive organs.

Disks that separate the bones of the spinal column (vertebrae) can sometimes slip or move out of place. If the disk damages or puts pressure on the cauda equina, a person may experience pain, numbness, or weakness in the lower extremities and back, known as cauda equina syndrome. Symptoms may also include bladder and bowel control problems and paralysis. Surgery to relieve the pressure on the cauda equina can reduce the symptoms.

See also Vertebral Column

Cell Body

The part of the neuron that contains organelles necessary for cell maintenance and survival; also called the soma. Within the cell body is the nucleus, an organelle that contains the genetic material (chromosomes) for cell development and synthesis of proteins. The nucleolus, located within the nucleus, produces ribosomes that help translate genetic information into proteins. Protein synthesis also occurs in the cell body within other groups of ribosomes called Nissl bodies. Ribosomes are also found on the endoplasmic reticulum, a system of tubes used to transport materials within a neuron.

After proteins are produced in a neuron, they are packaged by a membrane-bound structure called the Golgi apparatus. Microfilaments and neurotubules provide structural support to a neuron and also form a system to transport materials within a neuron. Mitochondria produce the energy necessary to fuel neuronal activities.

Cerebellum

Area of the brain above the brainstem and below the occipital lobe that is important for motor coordination, sensation, language, balance, and posture. The cerebellum weighs about 150 grams and has a volume similar to that of a tennis ball. Although the cerebellum makes up only 10% of the total brain volume, it contains about 80% of all the brain's neurons.[24]

The word *cerebellum* comes from Latin, meaning "little brain," and the appearance of this structure shares some features of the cerebral hemispheres. For example, both the cerebellum and cerebral hemispheres are divided into right and left sides and both brain structures are highly folded.

The cerebellum has two major divisions: the cerebellar cortex and the deep cerebellar nuclei. The cerebellar cortex contains several types of neurons, including granule cells, Purkinje cells, Golgi cells, stellate cells, and basket cells. Cells in the cerebellar cortex receive information from the cerebral cortex, spinal cord, and vestibular nuclei. The cerebellar nuclei are primarily involved with sending information out of the cerebellum to the brainstem and thalamus. This information eventually reaches the spinal cord and cerebral cortex.

Damage to the cerebellum may cause voluntary movement problems, tremors, and difficulty maintaining balance and posture. A few people have lived their entire lives missing most or all of their cerebellum.[25] Although most people born without a cerebellum have severe cognitive impairments, in rare cases, adults without a cerebellum show only mild to moderate symptoms. The absence of a significant disability in these people attests to the remarkable ability of the brain to adapt and compensate for brain injuries.

See also Brainstem; Cerebral Cortex; Occipital Lobe; Spinal Cord

Cerebral Cortex

Outermost layer of the cerebral hemispheres. Viewed from the top, the surface of the brain has the look of a giant walnut with folds called sulci (singular, sulcus) and bumps called gyri (singular, gyrus). The folds increase the amount of cerebral cortex that can fit within the skull. Just like fingerprints are unique for each person, so too is the pattern formed by the sulci and gyri of the cerebral cortex. Both sides of the cerebral cortex have the same general pattern of sulci and gyri, but the length, width, and shape of the sulci and gyri on the left and right sides of the brain can vary. The entire cerebral cortex forms a thin cap (1.5–4.5 mm thick) around most of the brain.

The cerebral cortex is divided into the three- or four-layered allocortex and the six-layered neocortex. The allocortex is a relatively older type of tissue than that of the neocortex, and is located in the middle part of the temporal lobes and involved with emotional behavior, olfaction, and memory. The neocortex makes up most of the cerebral cortex. Some areas of the neocortex process sensory information. For example, the visual cortex in the occipital lobe receives information about sight; the auditory cortex in the temporal lobe processes information about sound; the somatosensory cortex in the parietal lobe receives information about touch, pressure, and temperature from the skin. Motor areas of the neocortex in the frontal lobe are responsible for movement. Other parts of the neocortex, called association areas, integrate information from multiple brain regions and help with memory, decision making, attention, language, and other complex cognitive functions.

The word *cortex* comes from the Latin word for the bark of a tree because of the way this structure envelops most of the brain.

See also Frontal Lobe; Occipital Lobe; Parietal Lobe; Temporal Lobe

Cerebrospinal Fluid (CSF)

Clear fluid in the ventricular system of the brain and spinal cord. It may seem as if your brain fits solidly within your skull, but your brain (and spinal cord) actually floats in a clear, colorless liquid called cerebrospinal fluid (CSF). CSF also fills cavities inside the brain called ventricles.

Approximately 400–500 ml (1.5–2 cups) of CSF is produced every day in the lateral, third, and fourth ventricles of the brain by a structure called the choroid plexus. The CSF circulates through the brain by flowing through the ventricles. The fluid is absorbed into the bloodstream in the superior sagittal sinus through structures called arachnoid villi.

CSF acts to protect the brain and helps transport materials around the brain. By surrounding the brain, CSF acts as a buffer to cushion the brain after an impact to the head. CSF also reduces pressure at the base of brain because the brain floats in this layer of fluid. Because CSF flows in only one direction, the fluid can remove toxins and chemicals from the brain. The flow of CSF also serves to move hormones around the brain.

CSF can build up inside the ventricles if too much CSF is produced, if the ventricular system becomes blocked, or if CSF is not absorbed into the bloodstream properly. This can result in a condition called hydrocephalus, where the ventricles expand in size. Expansion of the ventricles can increase pressure within the skull and cause headaches, vision problems, cognitive difficulties, seizures, and poor coordination.

See also Meninges; Ventricles

Circle of Willis

Ring of blood vessels seen at the base of the brain. The circle of Willis is the brain's blood flow "fail-safe" mechanism; if one blood vessel in the pathway is blocked, it ensures that blood can still circulate. This pathway connects two major arteries—the internal carotid and vertebral arteries—that bring blood to the front and back of the brain. The internal carotid arteries supply blood to much of the cerebral cortex, and the vertebral arteries bring blood to the brainstem, cerebellum, occipital lobes, and part of the thalamus. The circle of Willis is completed by the basilar artery, anterior cerebral artery, anterior communicating artery, middle cerebral artery, posterior cerebral artery, and posterior communicating artery.

In 1664, while he was a professor of natural philosophy at Oxford University, Thomas Willis (1621–1675) wrote *Cerebri Anatome*. This work was based on Willis's dissections and observations of his patients' brains. *Cerebri Anatome* contains detailed descriptions (and illustrations by Sir Christopher Wren) of the cranial nerves and the ring of arteries at the base of the brain that is still known as the circle of Willis.

Thomas Willis was not the first to describe the circular blood vessel pattern at the base of the brain. Gabriel Fallopius (1523–1562) and Giulio Casserio (1552–1616) both noted the pattern, but their descriptions were incomplete. Willis provided a more complete description of the blood vessels and emphasized that the circular pattern was important, because if one artery was blocked, the ring of interconnecting blood vessels compensated for the blockage and allowed continued blood circulation.

See also Cranial Nerves

Cocaine

Local anesthetic and central nervous system stimulant derived from the *Erythroxylon coca* plant. Indigenous people in South America have known for thousands of years that chewing on coca leaves can help fight off fatigue. But it was not until 1860 when German chemist Albert Neiman (1834–1861) isolated cocaine, the active ingredient from the coca leaf. Soon after Neiman made his discovery, cocaine found its way into medicine and other products. In the early 1880s, Angelo Mariani (1838–1914) manufactured Vin Mariani, a "medicinal" wine with 11% alcohol and 6.5 mg of cocaine in every ounce. Austrian psychotherapist Sigmund Freud (1856–1939) also found a use for the drug when he prescribed cocaine for alcohol and morphine addictions. Unfortunately, Freud and many of his patients became addicted to cocaine. Cocaine found its way into various products, including Coca Cola, developed by John Pemberton (1831–1888) in 1886. The original recipe for Coca Cola called for the addition of cocaine and caffeine. Cocaine was removed from Coca Cola in 1906, but the drink still contains caffeine.

Depending on how cocaine is used (smoked, inhaled, or injected), the drug can reach the brain within a few seconds to a few minutes. In the brain, cocaine blocks the reuptake of the neurotransmitters dopamine, nor-epinephrine, and serotonin and causes the release of dopamine. Therefore, these neurotransmitters are available to act on receptors for an extended period of time. In the peripheral nervous system, cocaine can constrict blood vessels, dilate the pupils of the eye, and cause an irregular heartbeat. People who use cocaine may seem excited, have feelings of pleasure and well-being, and act with a great sense of confidence. Cocaine may also cause dizziness, headaches, anxiety, sleep problems, and hallucinations.

The cocaine high may last about an hour after which the person enters a period of depression. To get out of this depression, people may seek more cocaine and eventually become addicted. Without additional cocaine, people addicted to cocaine suffer withdrawal symptoms, such as feelings of depression, anxiousness, and paranoia.

Cocaine users face an elevated risk of a stroke because the drug increases blood pressure and constricts brain blood vessels. Some people who have overdosed on cocaine have suffered breathing and heart problems that resulted in their death. Celebrities whose deaths have been associated with cocaine use include basketball player Len Bias (1963–1986), comedian actors John Belushi (1949–1982) and Chris Farley (1964–1997), and singers Ike Turner (1931–2007) and Whitney Houston (1963–2012).

See also Caffeine; Neurotransmitters

Cochlea

Inner ear structure important for hearing. Sound is carried by changes in air pressure. The ridges and folds on your ears not only hold earrings and piercings, they also help direct sound waves into your ear. Sound waves travel to the end of the ear canal where they vibrate the eardrum (tympanic membrane). Vibrations of the eardrum set in motion the movement of three small bones (the ossicles) in the middle ear. The last of these bones (the stapes) moves another membrane called the oval window. Movement of the oval window pushes and pulls fluid inside the cochlea. The cochlea is where the vibration is changed into electrical signals that the brain uses to perceive sound.

The cochlea, which gets its name from the Greek and Latin words for "snail" because of their similar appearance, is filled with fluid and lined with another layer of

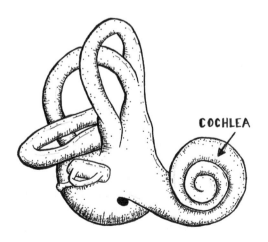

COCHLEA

tissue called the basilar membrane. The basilar membrane has special receptor cells (hair cells) that respond to movement. Movement of the fluid inside the cochlea moves the basilar membrane, which in turn stimulates hair cells. The hair cells generate electrical signals that move through the auditory nerve to the brain.

According to the World Health Organization, approximately 430 million people around the world have disabling hearing loss.[26] Some hearing loss is common as people age. This age-related hearing is often caused when hair cells in the inner ear are damaged or die. This type of hearing loss is permanent because hair cells do not regrow. The risk of hearing loss can be reduced by avoiding loud noises, including potentially damaging sounds at noisy events such as concerts and sports arenas. The volume of music played through earphones and headphones should also be turned down.

See also Ossicles

Computed Tomography (CT)

Medical imaging method that uses a series of X-ray beams passed through body tissues to create cross-sectional images of the brain and other body parts. CT images are collected by a computer that reassembles the pictures to form a three-dimensional image. A CT scan can be used to diagnose and locate brain tumors or skull fractures, guide surgeons during neurosurgical operations, and monitor the course of treatment.

The X-rays used in a CT scan do expose patients to low doses of ionizing radiation. However, exposure to radiation in a CT scan is kept to the minimum level required for the procedure. During some CT scans, patients are given a special dye (contrast agent) to produce images that better distinguish normal from abnormal tissue.

Much of the work toward the development of the CT scan was performed by Allan MacLeod Cormack (1924–1998) and Godfrey Newbold Hounsfield (1919–2004). Hounsfield's research was partially supported by funds generated by musicians, including the Beatles, who had signed with Electrical and Music Industries Limited (EMI).[27] Cormack and Hounsfield were awarded the Nobel Prize in Physiology or Medicine in 1979 for the invention of the CT scan.

Concussion

A traumatic brain injury caused when the brain strikes the inside of the skull. An impact to the head or shaking of the head can cause the brain to move and impact the skull. Although the brain is somewhat cushioned by a surrounding layer of cerebrospinal fluid, sudden movement of the head can force the brain into the skull and damage neurons and blood vessels. A person who has suffered a concussion will show a change in mental status and may experience dizziness, slurred speech, headache, confusion, sensitivity to light or sound, changes in mood, problems sleeping, and difficulties with memory. A person does not have to lose consciousness to have suffered a concussion.

The only way to determine if someone has had a concussion is by a clinical examination. During this examination, a doctor tests the person's sensory and cognitive abilities, such as hearing, vision, reflexes, balance, memory, and attention. The results of the exam help the doctor determine the severity of the concussion. For more severe concussions, doctors may order a brain imaging test, for example, magnetic resonance imaging (MRI) or computed tomography (CT) to check for brain damage.

Unfortunately, there are no medicines or drugs that can cure or heal a concussion. Instead, people who have had a concussion should seek medical care immediately. This includes people who have suffered a concussion while playing a sport: they should be removed from the game as soon as possible to prevent further injury. Players should not return to play until they are cleared by a medical professional. Concussion treatment starts with rest to help neurons recover and restore proper blood flow to affected areas; rest provides time for the brain to repair itself.[28]

To reduce the chances of sustaining a concussion, people should wear protective headgear when they ride a bike or motorcycle, ski, snowboard, skateboard, and roller-skate. Seat belts should also be worn by drivers and passengers while in a car, and young children should ride in appropriate car seats. Older adults should talk with their doctors about ways to reduce the risk of falls that can result in head injuries.

The word *concussion* comes from the Latin word *concutere*, meaning "to shake violently."

See also Cerebrospinal Fluid; Computed Tomography; Cranium; Magnetic Resonance Imaging

Congenital Insensitivity to Pain

Rare inherited condition rendering a person unable to perceive pain. Having a life free of physical pain might seem like a blessing. However, most people affected by congenital insensitivity to pain consider their condition a curse. Without the ability to feel pain, a person will not know when they break a bone, burn their skin, or bite their tongue. Internal injuries may also go undetected. Without treatment, these injuries can develop into severe, life-threatening infections and contribute to lower life expectancy.

People with congenital insensitivity to pain have normal perceptions to nonpainful touch and pressure, but they lack the ability to detect pain sensations. The underlying cause of the condition is a genetic mutation that results in a problem with voltage-gated sodium (Nav1.7) channels on neurons.[29] These channels are responsible for moving sodium ions across neuron membranes to generate action potentials. If a neuron that is normally used to send messages related to pain has a defective Nav1.7 channel, then it does not send a signal. Therefore, the brain never receives a message about pain.

A better understanding of pain and how Nav1.7 channels modulate pain may result in new treatments for people who suffer from chronic pain.

See also Action Potential

Coronavirus Disease 2019 (COVID-19)

Global disease caused by a severe acute respiratory syndrome coronavirus-2 (SARS-CoV-2) infection. COVID-19 infections may cause mild or severe illness; the most common symptoms of people infected with SARS-CoV-2 appear two to fourteen days after exposure to the virus and may include fever or chills, cough, difficulty breathing, and fatigue.

Although COVID-19 primarily affects the respiratory system, the virus appears to be responsible for problems involving other body systems, including the nervous system.[30, 31] For example, many people with mild or moderate COVID-19 symptoms may lose their sense of smell and taste. COVID-19 infections may also cause headaches, vision and auditory problems, muscular pain, and impaired consciousness. Even after respiratory problems due to an infection have cleared up, some people report lingering cognitive issues ("brain fog"), such as problems with attention, concentration, and memory that can last for several weeks or even months.

The route that the coronavirus takes to the brain is not yet confirmed, but there is some evidence that SARS-CoV-2 might access the brain by crossing the blood-brain barrier. After entering the brain, the virus may cause inflammation and damage blood vessels. These changes may result in a stroke and produce other neurological symptoms. A better understanding of how COVID-19 attacks the nervous system will help with the development of treatments for people affected by coronavirus disease.

See also Blood-Brain Barrier; Stroke

Corpus Callosum

Large collection of axons that connect the left and right hemispheres of the brain. Although not the only connection between the two sides of the brain, the corpus callosum is the largest link. Think of the corpus callosum as a neural superhighway with millions of lanes connecting each side of the brain. In fact, the

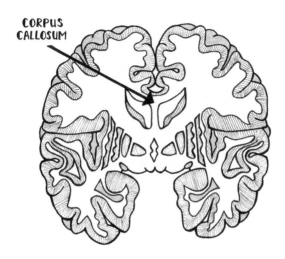

CORPUS
CALLOSUM

corpus callosum has approximately 200 million axons crossing from side to side to help coordinate sensory processes and higher cognitive functions within the cerebral cortex.

The importance of the corpus callosum in transferring information between the cerebral hemispheres was highlighted in the 1981 Nobel Prize–winning research of Roger Sperry (1913–1994). By studying people who underwent surgery that severed their corpus callosum, Sperry demonstrated that specific brain functions were dominant in either the right or left hemispheres. Additional insights about brain function and development have been provided by studying people who were born without a corpus callosum.[32] It may seem that being born without a corpus callosum (agenesis of the corpus callosum), which occurs in one in four thousand people, would result in a tremendous deficit. Some people with agenesis of the corpus callosum do have visual impairments, language difficulties, and problems with motor coordination, but other people born with this condition have few, if any, symptoms; they can go through life completely unaware that they are missing this interhemispheric connection. The lack of a corpus callosum is only discovered after these people undergo a brain scan. This observation suggests that the brain has significant redundant pathways for the transfer of information from one side to the other.

Although the corpus callosum is found in most mammals, marsupials (pouched mammals, such as the opossum) and monotremes (egg-laying mammals, such as the platypus) are born without this connection.[33] Instead, marsupials and monotremes have other pathways that connect the right and left sides of their brains.

See also Axon; Sperry, Roger Walcott

Cotard's Syndrome

Rare psychiatric disorder when individuals believe they are dead or have lost their blood or internal organs. In extreme cases, people with Cotard's syndrome claim that they have lost their soul or do not even exist. Even when presented with evidence that they are not dead or missing any organs, people with Cotard's syndrome will cling to their delusion.

Psychiatrists continue to debate whether Cotard's syndrome is a unique disorder or one that is secondary to other conditions, such as depression or anxiety.[34, 35] Symptoms of Cotard's syndrome have been reported in people with schizophrenia, dementia, migraine, Parkinson's disease, and multiple sclerosis. Brain damage to the cerebral cortex around the temporal and parietal lobes or cerebellum may also result in Cotard's syndrome beliefs.

Cotard's syndrome was described for the first time in 1880 by French psychiatrist and neurologist Jules Cotard (1840–1889) when he detailed the symptoms of a forty-three-year-old woman who believed her body was decomposing and that she did not have a brain, nerves, or internal organs. References to Cotard's syndrome have appeared in film. For example, in the 2008 movie *Synecdoche, New York*, the lead character played by Philip Seymour Hoffman (1967–2014) is aptly named Caden Cotard and has several themes related to death. On television, the zombies of the *Walking Dead* franchise might be based on Cotard's syndrome, although it is difficult to know what they (the zombies and the show writers) may be thinking.

Cranial Nerves

Twelve pairs of nerves that are connected directly to the brain. The cranial nerves send sensory information to the brain, control muscles, and help regulate internal organs, such as the heart, lungs, and stomach. Some of these nerves have sensory functions only (sensory), some only control muscles (motor), and some have both functions (mixed). By convention, each cranial nerve is identified by a Roman numeral:

   I   Olfactory nerve: Smell; sensory only
  II   Optic nerve: Vision; sensory only
 III   Oculomotor nerve: Eye movement and pupil constriction; motor only
  IV   Trochlear nerve: Eye movement; motor only

   V Trigeminal nerve: Touch and pain information from the face and head; control of muscles used for chewing; sensory and motor

  VI Abducens nerve: Eye movement; motor only

 VII Facial nerve: Taste from the front two-thirds of the tongue; touch and pain information from the ear; control of muscles used for facial expression; sensory and motor

VIII Vestibulocochlear nerve: Hearing and balance; sensory only

  IX Glossopharyngeal nerve: Taste from the back one-third of the tongue; touch and pain information from the tongue, tonsil, and pharynx; controls muscles used in swallowing; sensory and motor

   X Vagus nerve: Sensory information from and motor control of internal organs; sensory and motor

  XI Accessory nerve: Control of muscles used for head movement; motor only

 XII Hypoglossal nerve: Control of tongue muscles; motor only

A common memory device (mnemonic) to remember the names of the cranial nerves in order is **O**n **O**ld **O**lympus' **T**owering **T**op **A** **F**amous **V**ocal **G**erman **V**iewed **A** **H**op.

Cranium

The part of the skull that contains the brain. Some people call it a noggin, others call it a coconut, but whatever name is given to the skull, the brain is in it. The entire skull contains twenty-two bones of the head and face and three small bones (stapes, incus, and malleus) in each ear. The 3-pound (1.4 kg) adult human brain rests comfortably protected by eight bones of the skull that collectively are known as the cranium. These eight cranial bones include one frontal bone, two parietal bones, two temporal bones, one occipital bone, one sphenoid bone, and one ethmoid bone.

The skull has several large and small holes, called foramina, for blood vessels and nerves to enter and exit. All of the cranial nerves must pass through the cranium to reach the brain. For example, bundles of the olfactory nerve enter through foramina in the ethmoid bone, and the optic nerve from each eye projects through an opening in the sphenoid bone. The foramen magnum

is a large hole in the back of the occipital bone where the medulla ends and projects out of the skull to merge with the spinal cord.

The bones of the cranium join at sutures. In infants, these sutures have small gaps called fontanelles or "soft spots," which allow the brain to expand during development. Parents may be alarmed when they see their child's fontanelles pulsate as if the brain is moving in and out. Pulsation of the fontanelle is normal and the fontanelles close and become hard around the age of 1.5 years.

For centuries, the skull has been a symbol of death and mortality. In the eighteenth century, for example, the pirate known as Blackbeard flew a flag with skull and crossbones called the Jolly Roger. Today, a variation of the Jolly Roger is used as a warning that substances are poisonous or dangerous.

See also Cranial Nerves

Dendrite

Neuronal extensions that receive information from other neurons and take information to the cell body. Because of their branch-like appearance, dendrites were named for the Greek word meaning "tree." The extensive branching of dendrites is especially apparent in multipolar neurons.

The surfaces of dendrites are embedded with receptors that interact with neurotransmitters. When a neurotransmitter binds with a receptor, channels on the neuron open to allow the flow of ions (electrically charged particles). The flow of ions causes a small electrical signal to travel toward the neuron cell body. Unlike an action potential that travels in a neuron's axon, the electrical signals in dendrites degrade in size as they approach the cell body. These signals can be excitatory or inhibitory in nature, meaning that they can make the neuron more or less likely to cause an action potential.

A single neuron may receive input from thousands of other neurons through its dendrites. A neuron must integrate all of these signals, and only if the overall activity is above a set threshold will an action potential be generated.

See also Action Potential; Axon; Neuron; Neurotransmitters

Dopamine

Neurotransmitter found in the peripheral and central nervous system. Dopamine has gained the reputation of being the "feel-good" neurotransmitter because the release of this chemical in the brain is associated with feelings of pleasure and reward. But limiting dopamine's involvement to these functions does not do it justice; this neurotransmitter plays a critical role in movement, memory, learning, and arousal.[36]

Dopamine is derived from the amino acid tyrosine. In the presence of the enzyme tyrosine hydroxylase, tyrosine is converted to levodopa (L-DOPA). The enzyme aromatic l-amino acid decarboxylase (DOPA decarboxylase) converts L-DOPA to dopamine. The areas of the brain that are the main sources of dopamine are the substantia nigra, ventral tegmental area, and hypothalamus. These areas send axons to other parts of the brain (e.g.,cerebral cortex, striatum, hippocampus, amygdala, pituitary), where there are high concentrations of receptors for dopamine.

Alterations within the dopamine neurotransmitter system are associated with neurological disorders. For example, destruction of dopamine-producing neurons in the substantia nigra can result in Parkinson's disease. Dopamine does not cross the blood-brain barrier easily, so to relieve the symptoms of Parkinson's disease, patients can take L-DOPA, which crosses into the brain and converts to dopamine. Changes in dopamine (and other neurotransmitter) levels may also be responsible for psychiatric disorders, such as schizophrenia.[37]

The use of L-DOPA as a miracle cure for Parkinson's disease was popularized by the book *Awakenings*, written by neurologist Oliver Sacks (1933–2015) and published in 1973. The book describes a series of patients with a form of Parkinson's disease who were given L-DOPA and subsequently regained movement after a long period of immobility.[38] In 1990, *Awakenings* was made into a film starring Robin Williams (1951–2014) and Robert De Niro (1943–).

See also Parkinson's disease; Schizophrenia

Ecstasy (MDMA)

Synthetic psychoactive drug that alters mood and perception. Ecstasy (MDMA; 3,4-methyl-enedioxymethamphetamine) was initially developed in 1914 by the German drug company Merck as an appetite suppressant. In the 1970s, psychotherapists treated patients with Ecstasy as part of therapy to lower inhibitions and get people to open up and talk about their issues. Some studies suggest that Ecstasy can be used to treat posttraumatic stress disorder, depression, and anxiety disorders. Today in the US, MDMA is illegal, but it is used recreationally. The US Drug Enforcement Agency lists MDMA as a Schedule I drug, meaning that it has a high potential for abuse and has no currently accepted medical use.

In the brain, MDMA works to increase the release of serotonin, norepinephrine, and dopamine neurotransmitters or block the reuptake of these neurotransmitters. In both cases, the result is an increase in the availability of neurotransmitters within synapses. Increased activity of these neurochemicals leads to increased energy, accelerated heart rate, higher blood pressure, elevated mood, and reduced inhibitions. Many people who take MDMA also experience nausea, headaches, teeth clenching, increased sweating, and chills.

The addictive nature of MDMA is unclear. MDMA affects the same neurotransmitter systems as some addictive drugs, such as cocaine, and some experiments show that animals will take MDMA if given the opportunity (self-administration). However, few studies have investigated MDMA addiction and dependency in the general population.

MDMA is also known as Molly, Adam, XTC, and E.

See also Dopamine; Neurotransmitters; Serotonin; Synapse

Edwin Smith Surgical Papyrus

Ancient Egyptian text containing the oldest written record of the word *brain*. Named after the American Egyptologist who purchased the papyrus in 1862, the Edwin Smith Surgical Papyrus provides the first written accounts of the anatomy of the brain, the meninges (coverings of the brain), and cerebrospinal fluid. Written

around 1700 BC, the work is based on texts dating back to 3000 BC and may have been authored by Imhotep, an Egyptian physician. The papyrus was translated into English by James Henry in 1930.[39]

The papyrus is about 4.68 m long (15 ft, 3.5 in) and 32.5–33 cm wide (13 in) and contains descriptions of forty-eight medical cases. The patients described in these cases were likely injured by falls or suffered wounds in battle. Head injuries account for twenty-seven of the cases and the brain is mentioned seven times, but the word *nerve* is never used. Each case lists the type and location of an injury and how the patient should be examined, diagnosed, and treated. An injury that fractured the temporal bone of the skull reportedly caused the patient to lose his ability to speak. Therefore, the papyrus provides the first written description of aphasia.

The ancient Egyptians may have been the first people to write about the brain, but they did not hold the brain in high regard. When making mummies to prepare a body for the afterlife, the Egyptians scooped out the brain and discarded it.

See also Broca, Paul; Cerebrospinal Fluid; Meninges; Wernicke, Carl

Einstein's Brain

Mathematician and physicist Albert Einstein (1879–1955) suffered an abdominal aortic aneurysm and died at the age of seventy-six years on April 18, 1955. During Einstein's autopsy, Thomas S. Harvey (1912–2007), a pathologist at Princeton Hospital, removed Einstein's brain and kept it in a glass jar for many years.

Harvey slowly released his hold on the famous brain and provided small samples to laboratories for analysis. In 1985, the journal *Experimental Neurology* published a paper by Marian C. Diamond and her colleagues.[40] These researchers found one area of Einstein's brain (cerebral cortex area 39) had more glial cells for every neuron compared to this number in brains from a control group.

Other scientists have observed structural differences in glial cells, a thicker corpus callosum, a partially missing groove (sulcus), a thinner cerebral cortex (area 9), and a higher density of neurons in Einstein's brain.[41–43] These observations were used to suggest possible neuroanatomical links to Einstein's cognitive prowess.

The significance of the findings from these studies has been criticized because the brains used to compare to Einstein's brain may not have been adequate. For example, in one study, Einstein's brain was compared to the brains from people who were on average twelve years younger. Also, all of the studies had only one experimental subject: Einstein. Scientists do not know if other mathematical geniuses share the neuroanatomical features observed in Einstein's brain.

Einstein's great intellect was not reflected in a great brain size. Einstein's brain weighed only 1,230 gm compared to an average adult human brain weight of 1,400 gm.

See also Cerebral Cortex; Corpus Callosum; Glia

Electroencephalography (EEG)

Method used to record the electrical activity of the brain with scalp electrodes. When most people think of an EEG, they picture squiggly lines on a piece of paper or computer monitor that are generated from a person whose head has been connected by wires to a machine. This image of an EEG is correct: the record of brain activity (brain waves) is the result of the amplified electrical signals generated by neurons.

An EEG is typically recorded from electrodes placed on a person's scalp using the "10–20 system," which is based on the relative locations of different areas of

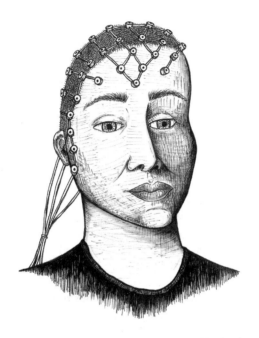

cerebral cortex under each electrode. Each electrode position has a letter to identify the lobe of the cortex (F = frontal lobe; T = temporal lobe; P = parietal lobe; O = occipital lobe; C = central) and a number to identify the hemispheric location. Even numbers (2, 4, 6, 8) refer to the right hemisphere and odd numbers (1, 3, 5, 7) refer to the left hemisphere. A *z* indicates that an electrode is on the midline. The "10" and "20" of the system refer to a 10% or 20% distance between each electrode.

When neurotransmitters cross a synapse and bind to receptors on a dendrite, they cause a small electrical signal. The sum of these small signals from thousands of neurons can be recorded from electrodes placed on the scalp. The EEG can be used to investigate altered states of consciousness, for example, during different stages of sleep and to help diagnose neurological disorders, such as epilepsy, sleep disorders, brain tumors, and stroke.

See also Action Potential; Epilepsy; Sleep; Stroke

Epilepsy

Neurological illness characterized by seizures. Imagine an electrical storm in the brain with neurons firing waves of abnormal signals. Now think of the possible consequences of this abnormal electrical activity: seizures, loss or alteration in consciousness, sensory changes, and convulsions. Although the signs and symptoms of epilepsy depend on the part of the brain affected and the speed that abnormal electrical activity spreads through the brain, a seizure is usually a serious health problem.

Partial seizures and generalized seizures are the primary symptoms of epilepsy. Partial seizures involve only a small part of the brain and may (complex) or may not (simple) affect consciousness. Generalized seizures affect both sides of the brain and usually cause changes in consciousness.

Most cases of epilepsy have no known cause. For other cases, head injury, brain tumors, stroke, or infections can cause the disorder. Seizures can sometimes be triggered by stressful events, poor sleep, bright lights, loud sounds, and low blood sugar.

Antiepileptic drugs, such as carbamazepine and phenytoin, are successful in treating many people with epilepsy by reducing the abnormal firing of neurons in the cerebral cortex. In cases when medication does not work to control seizures, a person may have surgery to remove the brain tissue where the seizure starts. Other surgical treatments include a corpus callosotomy (split-brain surgery) that cuts the corpus callosum to prevent the spread of abnormal signals from one side of the brain to the other and a hemispherectomy that removes a cerebral hemisphere.

Famous people who have been diagnosed with or are presumed to have had epilepsy include writers Fyodor Dostoyevsky (1821–1881), Truman Capote (1924–1984), Edgar Allan Poe (1809–1849), and Lewis Carroll (1832–1898); actors Richard Burton (1925–1984), Margaux Hemingway (1954–1996), Hugo Weaving (1960–), and Danny Glover (1946–); composer George Gershwin (1898–1937); singers Susan Boyle (1961–), Neil Young (1945–), Adam Horovitz (1966–), Elton John (1947–), Lil Wayne (1982–), and Prince (1958–2016); athlete Florence Griffith Joyner (1959–1998); Russian politician Vladimir Lenin (1870–1924); and French politician Napoleon Bonaparte (1769–1821). In 1979, Anthony (Tony) Coelho (1942–) became the first person with epilepsy to be elected to the US House of Representatives.

Fetal Alcohol Syndrome

Fetal abnormalities caused by consumption of alcohol by pregnant women. When a pregnant woman drinks an alcoholic beverage, she is exposing not only herself to the effects of alcohol but also her developing fetus. As a lipid-soluble molecule, alcohol moves easily across the placenta from a mother's blood supply to that of her developing fetus. Fetal exposure to alcohol can have devastating effects on physical and cognitive development that can last a lifetime.

Alcohol can disrupt normal brain development profoundly. For example, the development of the corpus callosum may be impaired, the size of the basal ganglia may be reduced, and damage to the cerebellum, hippocampus, and cerebral cortex may occur. Babies born with fetal alcohol syndrome may also have smaller heads, poor coordination, hyperactivity, and abnormal facial features (narrow eye openings, thin lips, and a smooth area above the upper lip). Children born with fetal alcohol syndrome may have long-lasting learning disabilities and problems with attention, decision making, controlling impulses, and communicating.

The type of alcohol consumed—beer, wine, spirits—makes no difference to the developing brain. There is no known safe level of alcohol for the developing fetus.

See also Alcohol; Cerebellum; Cerebral Cortex; Corpus Callosum; Hippocampus

Fregoli Syndrome

Rare psychiatric disorder characterized by the delusional belief that other people can change their appearance or shape. Think of quick-change artists who can alter their appearance with a few twists and twirls and disguise themselves easily. People with Fregoli syndrome often believe that they are being persecuted by the person in disguise.[44]

Capgras syndrome and Fregoli syndrome are similar in that they both involve abnormal perception and identification of faces. However, in Capgras syndrome, a person believes that someone familiar to them has been replaced by an impostor. In Fregoli syndrome, delusions involve the belief that another person can assume the identities of several other people known to the patient. For example, a person with Fregoli syndrome may believe that one person is pretending to be a bus driver at one time and then a librarian at another time.

Fregoli syndrome is named after Italian actor Leopoldo (Luigi) Fregoli (1867–1936), a famous quick-change artist, entertainer, and impersonator of his time.

See also Capgras Syndrome

Frontal Lobe

Part of each cerebral hemisphere located in front of the parietal lobe; responsible for processing information about emotions, planning, memory, problem solving, and movement. Rosemary Kennedy (1918–2005), sister of former US president John F. Kennedy (1917–1963), was prone to severe mood swings and violent outbursts. Rosemary's father had heard of a new brain surgery that might control Rosemary's behavior. When Rosemary was twenty-three years old, her father gave doctors permission to perform the operation on his daughter.[45] During the operation, neurosurgeon James W. Watts (1904–1994), with assistance from psychiatrist Walter Freeman (1895–1972), plunged a metal probe shaped like a butter knife into Rosemary's brain to separate the prefrontal lobe from the remaining frontal lobe. The operation was a failure and Rosemary was left unable to speak, walk, and take care of herself.

The devastating consequences of Rosemary Kennedy's operation (a lobotomy or leucotomy) illustrate the importance of the frontal lobes to higher cognitive functions, emotions, and personality. The frontal lobes also contain the motor cortex, a strip of tissue that controls movement of all parts of the body. Like the sensory cortex in the parietal lobe, the motor cortex contains a "map" of the body, with larger amounts of brain tissue devoted to body areas with better dexterity (e.g., fingers, hands, mouth). Broca's area, the area of cortex responsible for the production of speech, is also located in the frontal lobes.

Portuguese physician António Egas Moniz (1874–1955) performed some of the first prefrontal lobotomies in humans during the mid-1930s. In his first cases, Moniz injected alcohol into the frontal lobes to destroy brain tissue. Later, Moniz used a wire loop to sever the connections between the prefrontal cortex and the rest of the brain. Moniz won the Nobel Prize for Physiology or Medicine in 1949 for this work, but because of the damage done by the procedure, families of patients have tried (unsuccessfully) to have the prize revoked.

See also Broca, Paul; Gage, Phineas; Parietal Lobe

G ABA (Gamma-aminobutyric acid)

Common neurotransmitter in the brain and spinal cord that inhibits the generation of an action potential. When GABA binds to receptors on postsynaptic GABA receptors, it works to hyperpolarize a neuron and inhibit the transmission of an action potential. GABA can bind with two types of receptors to hyperpolarize a neuron: (1) binding to GABA-A receptors results in the flow of chloride ions into a neuron; and (2) binding to GABA-B receptors increases the conductance of potassium and inhibits calcium channels on presynaptic neurons, slowing neurotransmitter release. In either case, the electrical potential inside a neuron becomes more negative relative to the outside of a neuron, making it more difficult for an action potential to be generated.

GABA is one of the most abundant neurotransmitters in the mammalian brain and is used in multiple neural circuits in the central nervous system. Disruptions in the GABA neurotransmitter system have been linked to neurologic and psychiatric disorders, such as Huntington's disease, Alzheimer's disease, schizophrenia, depression, seizures, dystonia, and spasticity.[46]

See also Action Potential; Neurotransmitters; Synapse

Gage, Phineas (1823–1860)

Railroad foreman who had an unfortunate job accident that turned him into one of the most famous figures in the history of brain research. In Cavendish, Vermont, on September 13, 1848, Phineas Gage filled a small hole with gunpowder while he worked on the railroad. When Gage compacted the powder with an iron rod, a spark was produced. The spark ignited the gunpowder and sent the 3-foot, 7-inch, iron rod up through his cheekbone, through the left frontal lobe of his brain, and out the top of his skull. The accident also propelled Phineas Gage into neuroscience textbooks around the world.

Gage never lost consciousness after the accident and was taken to a doctor who did his best to repair the wounds to Gage's head. A subsequent brain infection sent Gage into a coma for a few days, but he recovered. However, Gage did lose sight in his left eye. As Gage recovered, people remarked about how his personality had

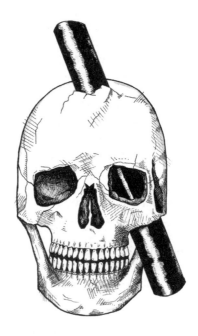

changed. Once a respected supervisor, Gage was now impulsive, rude, and discourteous. Because of his change in behavior, Gage was not rehired to work on the railroad.

Eventually, Gage found work as a stagecoach driver and was able hold down a job.[47] The brain injury suffered by Gage provided scientists with evidence for the importance of the frontal lobe in decision making, planning, personality, and problem solving and also for the brain's ability to repair itself.

Gage's skull and tamping iron were donated by Gage's physician, John Harlow (1819–1907), to the Warren Anatomical Museum at Harvard University in 1868.

Galen (130?–216?)

Greek physician and philosopher. Although Galen was born in present-day Turkey, he traveled to Greece and Egypt to study medicine. Galen worked as a doctor for gladiators, and as luck would have it (at least for Galen), he gained significant experience in the medical arts and knowledge of the human body by tending to the wounds of injured fighters. Working for Roman emperors, Galen was able to conduct medical research and publish his work.

Galen wrote extensively about anatomy, including descriptions of the nervous system. Because human dissection was forbidden, Galen performed dissections on animals and assumed that what he observed would also be true for humans. Galen's ideas about anatomy and physiology went unchallenged for more than a thousand years until scientists during the Renaissance began to examine human remains. It turns out that many of Galen's ideas about the human body were incorrect. For example, Galen agreed with Aristotle that the brain was not the seat of consciousness.

See also Aristotle; Vesalius, Andreas

Galvani, Luigi (1737–1798)

Italian physician who studied the effects of electricity on muscles and pioneered the field of electrophysiology. Galvani's early scientific interests focused on bones and bird anatomy. Galvani's claim to fame started when someone in his lab accidentally shocked a frog nerve with a scalpel that was connected to an electric generator. This accidental shock caused the frog leg to twitch and launched new experiments that contributed to Galvani's theory of "animal electricity."

Results of Galvani's experiments were detailed in a 1791 publication titled *De viribus electricitatis in motu musculari commentarius* (*Commentary on the Effect of Electricity on Muscular Motion*). Galvani demonstrated that when a frog's nerve was touched with different metals, it would cause the frog's muscles to contract. He noted that both natural electricity (lightning) and artificial electricity (friction) could cause muscles to twitch. These observations led him to conclude that animal tissue had its own internal force or "animal electricity." Galvani thought incorrectly that these experiments supported the notion that animals have a fluid similar to electricity and that the metals somehow released the electricity.

Alessandro Volta (1745–1827) became Galvani's main critic; he argued that there was no animal electrical fluid but asserted instead that it was contact with two different metals that resulted in the electricity that caused muscles to twitch.[48, 49] Although Galvani was wrong about animal electrical fluids, his observations laid the foundation of nerve/muscle physiology. In a way, Galvani was ahead of his time: nerves and muscles do generate electrical impulses (e.g., action potentials).

Galvani's later years were burdened with troubles in his personal life. Lucia Galeazzi Galvani, Galvani's wife and research collaborator, died in 1788, and the French Revolution prevented Galvani from continuing his experiments. When Napoleon Bonaparte's army occupied Northern Italy, university faculty members were required to take an oath of allegiance to the republic. Galvani refused to take the oath and lost his job.

See also Action Potential; Volta, Alessandro

Glia

Nonneural support cells of the nervous system. Glial cells (glia) do not get as much attention as the more well-known nerve cell (neuron). Unlike neurons, glial cells do not generate action potentials. Even the origin of the word *glia*, from the Greek word meaning "glue," was coined by early scientists because glial cells were thought to function only to hold the nervous system together. Today, we know that glial cells play essential roles in nervous system function.[50]

Several types of glia are found in the central and peripheral nervous systems. Astrocytes are star-shaped glial cells that remove brain debris, move nutrients to neurons, provide structural support for neurons, and regulate neurotransmitter levels around neurons. Satellite cells in the peripheral nervous system provide structural support to neurons. Microglia work like astrocytes by digesting dead neurons and removing toxins. Microglia may also help shape the developing brain by pruning unneeded synapses. In the central nervous system, oligodendroglia form myelin that wraps about neuronal axons. In the peripheral nervous system, the job of producing myelin is taken up by Schwann cells.

Neurons might be considered the celebrities of the nervous system, but without glia, the show would not go on.

See also Action Potential; Neurotransmitters; Neuron; Myelin

Golgi, Camillo (1843–1926)

Italian neuroscientist who discovered an innovative method capable of staining an entire neuron. After

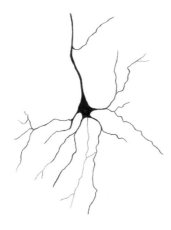

graduating from medical school at the University of Pavia, Golgi concerned himself primarily with psychiatry. However, when he set up a small laboratory in a hospital kitchen, Golgi turned his attention to the organization of the nervous system.

While working in his kitchen lab, Golgi discovered that a chemical procedure using silver nitrate and potassium dichromate stained individual neurons in their entirety. Golgi named this process the "black reaction" because of the color of the stained tissue.[51, 52] Golgi's observations of the tissue led him to the idea that the nervous system was constructed like a large network of connected neurons. This "reticular theory" was in direct opposition to Santiago Ramón y Cajal's "neuron theory" in which the nervous system was composed of individual cells. The arrival of better neuroanatomical methods proved Ramón y Cajal's neuron theory correct.

Several structures are named after Golgi, including Golgi I cells, Golgi II cells, Golgi body, and the Golgi tendon organ. In 1906, Golgi shared the Nobel Prize in Physiology or Medicine with his rival Ramón y Cajal for his contributions to understanding the structure of the nervous system.

See also Neuron; Ramón y Cajal, Santiago

H ippocampus

Brain structure located behind the amygdala in the temporal lobe; important for the formation of memories. When early neuroanatomists dissected the middle part of the temporal lobe, they saw a structure that resembled a white silkworm or a ram's horn. To Italian surgeon and anatomist Julius Caesar Arantius (1530–1589), the structure looked like a seahorse, so he named it the hippocampus, the Latin word for this animal.[53]

The hippocampus is essential for getting information into long-term memory. More specifically, the hippocampus helps build memories dealing with facts, names, and events (declarative or explicit memories). People with damage to the hippocampus develop anterograde amnesia because they cannot form new declarative memories. Memories established before hippocampal damage remain intact, but new declarative memories cannot be made.

In addition to its role in memory, the hippocampus may be viewed as the brain's own global positioning system (GPS), important for forming maps of an organism's environment. Scientists have discovered hippocampal neurons called place cells that respond when animals are in a particular location. Another type of neuron, a grid cell, is found in brain areas adjacent to the hippocampus. Grid cells are activated by a variety of evenly distributed positions that create a tiled map of the space. Together place cells and grid cells assist in helping an animal know where they are in space. Interestingly, London taxi drivers who successfully passed their licensing exam requiring memorization of thousands of streets and locations had larger hippocampi than drivers who failed the exam.[54] This observation provides further evidence to the importance of the hippocampus in spatial navigation.

John O'Keefe (1939–), May-Britt Moser (1963–), and Edvard I. Moser (1962–) shared the 2014 Nobel Prize in Physiology or Medicine for their studies of the hippocampus and the brain's positioning system.

See also Amygdala; Molaison, Henry; Temporal Lobe

Huntington's Disease

Fatal genetic neurodegenerative disorder characterized by abnormal movements and progressive dementia. Huntington's disease is an inherited neurological disease that affects approximately forty thousand people in the United States. The child of a parent who has the disease has a 50% chance of having the disorder.

The specific cause of Huntington's disease is a mutation in the gene for a protein called huntingtin. This mutation results in an abnormal repetition of the several DNA building blocks that damage neurons. Huntington's disease damages neurons in basal ganglia areas of the brain, especially in the caudate nucleus and globus pallidus.[55] Injury to these neurons may cause dementia, chorea (jerky, random movements of the body), poor coordination, depression, memory loss, and mood swings. These symptoms usually appear when a person is between thirty and fifty years old. Although a few drugs can reduce the symptoms of the disorder, there is no cure for Huntington's disease and patients usually die within ten to twenty-five years of the onset of symptoms.

Folk singer Woody Guthrie (1912–1967) inherited Huntington's disease from his mother.[56] Guthrie's health deteriorated to the point where he could not talk and lost control of voluntary movement. Two of Woody's children also had Huntington's disease, although his singer-songwriter son, Arlo Guthrie (1947–), did not inherit the disorder.

WOODY GUTHRIE

Lead
　　Heavy metal element that is toxic to the nervous system and other body systems. For thousands of years, lead has been a valuable product used in pipes, cooking pots, cosmetics, glazes, and paints. Although lead was removed from gasoline and house paint in the late twentieth century in the US, this heavy metal still exists in homes with old paint and in buildings with old pipes. The US Department of Health and Human Services estimates that thirty-seven million homes in the US contain lead-based paint.[57]

Exposure to lead poses a serious health risk, especially to children whose brains are still developing. Children may inhale lead dust or eat old paint chips that have peeled off from a wall. Lead can interfere with how neurons are connected and how neurons communicate with each other. Disruption in neural pathways can have devastating behavioral and neurological effects, including problems with intelligence, memory, attention, movement, and mood.[58] Handwashing, especially before eating, may help reduce the accidental ingestion of lead. Anyone worried that they may be exposed to lead can have their blood tested for lead levels.

The chemical symbol for lead is Pb from the Latin word *plumbum*, meaning "waterworks." As you might guess, the English words *plumber* and *plumbing* are derived from this Latin word.

Levi-Montalcini, Rita (1909–2012)

Italian-American neuroscientist who made pioneering discoveries about nerve growth factor. Levi-Montalcini pursued her scientific career during turbulent times.[59]

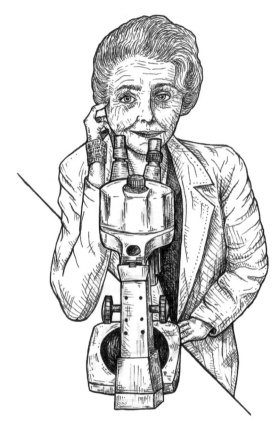

RITA LEVI-MONTALCINI

Born in Turin, Italy, into a wealthy Jewish family, Levi-Montalcini studied medicine at the University of Turin, where she was mentored by several scientists including Salvador Edward Luria (1912–1991) and Renato

Dulbecco (1914–2012), who both later won Nobel Prizes. A few years after Levi-Montalcini graduated from the University of Turin, the Fascist government of Benito Mussolini (1883–1945) stripped Jews of their Italian citizenship and prohibited non-Aryan Italian citizens from holding professional careers. This proclamation forced Levi-Montalcini to leave Italy for Brussels, Belgium.

Levi-Montalcini returned to Turin a few weeks before German forces invaded Belgium. Back in Turin, she first built a small laboratory in her bedroom, and when bombing by Allied Forces intensified, she built another home lab further away from the city. When the German army invaded Italy in 1943, Levi-Montalcini moved to Florence, where she spent the rest of the war living underground and providing medical care to war refugees.

After World War II, Levi-Montalcini accepted a position to work with Viktor Hamburger (1900–2001) at Washington University (St. Louis, MO). While at Washington University, Levi-Montalcini studied neural development and the proteins and amino acids that help neurons grow. Levi-Montalcini's discovery of nerve growth factor, a protein that stimulates neuronal growth, has transformed our understanding of how the nervous system develops and provides a potential pathway toward therapies to repair neurons damaged by neurodegenerative diseases.

In recognition of her discovery of nerve growth factor, Levi-Montalcini was awarded the Nobel Prize for Physiology or Medicine in 1986. Later in life, Levi-Montalcini was active in politics and in 2001 she was appointed as a senator for life by the president of Italy.

Lewy Body Dementia

Progressive neurodegenerative disease characterized by widespread Lewy bodies in the brain, resulting in a decline in cognitive abilities. In the months before his death, comedian Robin Williams (1951–2014) was plagued with depression, anxiety, sleep problems, memory loss, paranoia, and delusions. On August 11, 2014, Williams took his own life. An autopsy revealed that deposits of Lewy bodies had spread throughout Williams' brain and confirmed a diagnosis of Lewy body dementia.

People with Lewy body dementia have abnormal amounts of a protein called alpha-synuclein in their brains. Lewy bodies are composed of these proteins. Alpha-synuclein is found in healthy brains, where it helps with neurotransmission. The problem with alpha-synuclein is when it collects inside neurons, especially those neurons that use the neurotransmitters acetylcholine and dopamine, and prevents neurons from working properly. Eventually, the affected neurons die, causing Lewy body dementia. Lewy bodies often target the cerebral cortex, hippocampus, basal ganglia, and brainstem. This widespread damage accounts for the problems with perception, sleep, thinking, memory, emotions, movement, and language seen in people with Lewy body dementia.

Unfortunately, the cause of the abnormal collection of Lewy bodies inside neurons is not known and there are no cures for the disorder. Some drugs can manage the symptoms of Lewy body dementia (e.g., medications to improve sleep or antidepressants to treat depression), but there are no therapies to halt the progression of the disease.

Lewy bodies are named after Frederich H. Lewy (1885–1950), a German-American neurologist who discovered the abnormal proteins in 1912.[60]

See also Neurotransmitters

Loewi, Otto (1873–1961)

German-American neuroscientist responsible for the discovery of the neurotransmitter acetylcholine. One night in the early 1920s, Loewi woke from a dream with an idea for a new experiment. He wrote down his thoughts and then went back to sleep. Unfortunately, the next morning he could not read his writing or remember the fantastic experiment. Fortunately for Loewi and the field of neuroscience, the scientist had the same dream the next night and, when he awoke, he went directly to his lab.

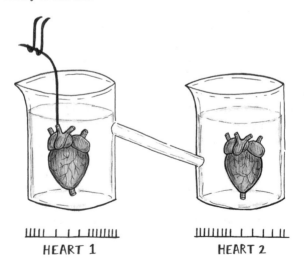

HEART 1   HEART 2

In the lab, Loewi put a frog heart into a container with fluid. When Loewi shocked the vagus nerve of the heart, he noticed that the heartbeat slowed down.[61] When the fluid from the container bathed a different frog heart in another container, Loewi saw that this second frog heart also slowed. Loewi surmised that the vagus nerve somehow released a chemical into the fluid that affected both hearts. Loewi was correct: he had discovered the neurotransmitter acetylcholine, which he named Vagusstoff.

Loewi and Henry Dale (1876–1968) shared the 1936 Nobel Prize for Physiology or Medicine for their discoveries about chemical transmission of nerve impulses.

See also Neurotransmitters

Lysergic Acid Diethylamide (LSD)

Hallucinogenic drug that alters perception, mood, and sense of reality. On one late afternoon in April, 1943, Swiss chemist Albert Hofmann (1906–2008) decided to try a small dose of a new drug he created.[62] On his bicycle ride home after taking the drug, Hofmann experienced visual hallucinations, feelings of paralysis, and dizziness. The cause of these strange symptoms? Hofmann had dosed himself with lysergic acid diethylamide (LSD).

LSD has no color, no odor, and no taste, but amounts as small as 0.05 mg can produce hallucinations. The effects of LSD vary depending on the dose and with the expectations and mood of the user. After taking LSD, a person may begin to feel the effects of the drug in 30 to 60 minutes and the "trip" can last 12 hours. Common effects include visual hallucinations, anxiety, altered emotions, increased heart rate and body temperature, and distortions in the perception of time and space.

Although all of the brain mechanisms responsible for the effects of LSD are not known, it is likely that brain pathways that use the neurotransmitter serotonin are involved. Serotonin plays important roles in mood, motor control, and regulation of body temperature. The chemical structure of LSD is similar to that of the neurotransmitter serotonin, so the effects of LSD are likely caused by activation of neurons that use serotonin.

In the US, LSD is now listed as a Schedule I drug (no medical use and a high potential for abuse), with a penalty of five years in prison for possession of 1 gm.[63] However, in the 1950s and 1960s, many studies investigated the possible therapeutic uses of LSD, especially for the treatment of alcoholism. Within the past ten years, there has been a renewed interest in LSD-assisted psychotherapy for anxiety, depression, and pain.[64, 65] Additional research is needed to determine if LSD can be used safely and effectively to treat neurological and mental disorders.

See also Mushrooms, Hallucinogenic; Neurotransmitters; Serotonin

Magnetic Resonance Imaging (MRI)

Noninvasive method to provide an anatomical view of the brain by detecting the displacement of radio frequency signals in a magnetic field. It is best that people are not claustrophobic when getting into an MRI machine because they are inserted into a narrow tube and must remain still. After a person is placed in an MRI machine, magnets are turned on to align protons within the person's body tissues. Radio waves are then sent through the tissue to stimulate the

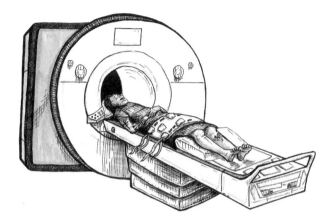

protons and change their spin. When the radio wave
signal is turned off, proton energy is detected by the
machine as the protons realign. The final image is created because different tissues respond in different ways
to the radio signal within the magnetic field.

MRI is used to detect and diagnose neurological disorders because healthy and diseased tissue can be seen
on the final image. MRI is also safer than other brain imaging methods because it does not use X-rays or radioactive material. However, patients must remove all metal
objects before they undergo MRI testing because of the
strong magnetic pull of the machine.

See also Computed Tomography

Marijuana
Dried leaves, flowers, stems, and seeds from the *Cannabis sativa* or *Cannabis indica* plants. Once an illegal drug,
marijuana is now legal in many states for medicinal and

recreational purposes. Marijuana contains several mind-altering chemicals, but its most well-known, mind-altering component is delta-9-tetrahydrocannabinol (THC). Cannabidiol (CBD), a nonpsychoactive component of marijuana, has received recent attention for its possible health benefits as a treatment for pain, nausea, and epilepsy. Marijuana is most commonly rolled and smoked, but it can also be cooked into baked goods, brewed into a tea, or mixed into other food such as candy ("edibles"). THC has also found its way into vaping products.

THC acts on the brain's endocannabinoid neurotransmitter system. Neurons that are part of the endocannabinoid system have receptors that are activated by THC. These receptors are especially numerous in the hippocampus, cerebral cortex, cerebellum, and basal ganglia, which are brain areas important for functions such as memory, decision making, perception, and movement. Therefore, stimulation of these areas by THC can cause relaxation, altered perception, reduced attention, lack of coordination, and disorientation.

Although there are no documented cases of anyone overdosing and dying after using marijuana, the drug can cause undesirable side effects, such as anxiety, paranoia, and even psychosis. Also, according to the National Institute on Drug Abuse, approximately 30% of people who use marijuana develop a marijuana use disorder (addiction and dependence).[66]

See also Neurotransmitters

Meninges

Series of three membranes (dura mater, arachnoid, pia mater) that cover the brain and spinal cord. The meninges provide a protective barrier between the brain and skull and between the spinal cord and vertebrae. The outermost layer of the meninges, just below the bones of the skull, is called the dura mater. The name comes from the Latin words meaning "tough mother" because this tissue is thick and strong. The dura helps keep the brain in place within the skull and protects the brain from movements that may cause injury. Below the dura is the middle layer of the meninges, called the arachnoid. The spider web–like arachnoid provides a channel for the circulation of cerebrospinal fluid. The inner layer of the meninges, the one closest to the brain, is called the pia mater. The pia mater (from Latin, meaning "soft mother") is very thin and transparent.

Meningitis is a serious health concern when the meninges are inflamed by bacteria, viruses, or less often by fungi. Common symptoms of meningitis include fever, headache, aversion to bright light, and stiff neck. A sample of cerebrospinal fluid can identify the type of meningitis and help determine treatment (e.g., antibiotics or antifungal medications). Vaccines for some forms of bacterial meningitis are effective and recommended for high-risk populations, such as university students who live in dormitories. Left untreated, bacterial meningitis can be fatal.

An easy way to remember the order of the meninges is to use the mnemonic "the meninges **PAD** the brain," where **P** = pia, **A** = arachnoid, and **D** = dura.

See also Cerebrospinal Fluid; Cranium; Vertebral Column

Mercury

Heavy metal element with neurotoxic actions. In his book *Alice's Adventures in Wonderland* (1865), Lewis Carroll (1832–1898) described the eccentric behavior of the Hatter, a hatmaker. One theory is that Carroll's "mad hatter" was based on people who made felt hats in the nineteenth century. These hatmakers were often exposed to neurotoxic levels of mercury during the felting process. Mercury poisoning can result in personality, mood, and memory changes. President Abraham Lincoln showed signs of mad hatter syndrome that were likely tied to his consumption of mercury in pills ("blue mass") he used to treat depression.[67] Fortunately, Lincoln stopped taking the pills soon after his presidential inauguration, and his behavior stabilized.

The ingestion, inhalation, and absorption of mercury can have devastating consequences on the nervous system, especially on developing brains. Mercury targets the cerebral cortex, cerebellum, spinal cord, and peripheral nerves by killing neurons and interfering with neurotransmission. Perhaps the most well-known environmental disaster caused by mercury occurred during the 1950s when water polluted with methylmercury was dumped into Minamata Bay, Japan. Many children born to mothers who ate contaminated seafood developed neurological problems, including paralysis, seizures, speech disorders, and sensory (hearing, visual, touch) disorders.

In case of a mercury spill, for example, from a broken, old thermometer, the mercury should be scooped up carefully with a piece of paper or cardboard and placed in a sealed bottle. The bottle should be disposed with other hazardous waste, and anyone handling the mercury should wash their hands. A vacuum cleaner should never be used to pick up mercury because it will just contaminate the vacuum cleaner and spread mercury vapors into the air for everyone to breathe.

See also Lead; Neurotoxin

Milner, Brenda (1918–)

British-Canadian neuropsychologist. When Milner enrolled at Cambridge University in 1936, she studied mathematics before switching to psychology. As an undergraduate student, Milner developed an interest in how studying people with brain damage could provide insights about normal cognitive function. She continued her academic studies at Cambridge and in 1949 received her master's degree in experimental psychology.

Under the supervision of Donald Hebb (1904–1985), Milner received her PhD from the Department of Psychology at McGill University in 1952 for her work on the effects of temporal lobe damage on intelligence and memory. Milner then joined the lab of neurosurgeon Wilder Penfield (1891–1976). In April 1955, Milner met Henry Molaison (1926–2008), a patient who would change her life and establish her as one of the world's most influential neuroscientists.

Molaison had temporal lobe surgery to relieve his epileptic seizures. Although the surgery to treat the seizures was successful, Molaison was never again able to form new memories. Milner worked with Molaison for more than thirty years to understand why Molaison could learn motor tasks but not remember how he learned them. Milner's careful experiments showed that the temporal lobes play an essential role in converting short-term memories into long-term memories and that the brain has multiple memory systems. Milner's more recent research focuses on brain mechanisms important for spatial memory and language.

Although Milner is now more than one hundred years old, she maintains her appointment at McGill University in Montreal, Canada. Milner has received many professional honors and awards; in 2021, the Cognitive Neuroscience Society held a special symposium to honor Milner and her accomplishments.

See also Molaison, Henry; Penfield, Wilder; Temporal Lobe

Molaison, Henry (1926–2008)

Patient who developed memory problems after undergoing brain surgery to control seizures. Like Phineas Gage, Henry Molaison (also known as HM) has found his way into neuroscience textbooks as a famous patient whom all students of neuroscience should know.

When Henry Molaison was seven years old, he was hit by a bicycle and developed seizures a few years later. As Henry grew, his seizures became more severe. When Henry was twenty-seven years old, he underwent brain surgery to control the seizures. During this operation, neurosurgeons removed the hippocampus, amygdala, and parts of the temporal cortex on both sides of Molaison's brain. Although the surgery reduced Molaison's seizures, Henry was left with a strange memory disorder: he could no longer form new memories. He could remember things that happened before the surgery, but it was as if he was frozen in time because he could not make new memories for the remainder of his life.[68, 69]

Although Molaison could not remember new facts, people, or events, he could learn new motor skills. For example, Molaison could retain his ability to "mirror draw," but he could not recall how he learned the task. This observation and other experiments showed that declarative memories (e.g., facts, events, dates) and procedural memories (e.g., skills) are retained in different parts of the brain.

Molaison died on December 2, 2008; a year and two days later, his brain was sliced into thin sections by researchers at the University of California, San Diego.[70] An atlas of Molaison's brain can be viewed online at the Brain Observatory.[71]

See also Gage, Phineas; Hippocampus; Milner, Brenda

Multiple Sclerosis

Neurological disease caused by damage to a neuron's myelin sheath. Myelin is a fatty substance that insulates neurons and helps speed electrical signals as they travel down an axon. Multiple sclerosis is an autoimmune disease because a person's immune system turns against its own body. By attacking myelin in the central nervous system, multiple sclerosis results in inflammation and damage that disrupts the travel of information within neurons.

Environmental and genetic variables appear to contribute to multiple sclerosis, but the actual cause of the disease is not known. People with multiple sclerosis may experience pain, fatigue, numbness or tingling in their hands and feet, difficulty walking and keeping balance, cognitive changes, and visual problems. In some people the symptoms continually worsen, while in other people the symptoms come and go. To diagnose multiple

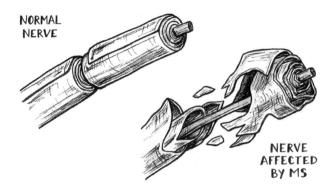

NORMAL
NERVE

NERVE
AFFECTED
BY MS

sclerosis, magnetic resonance imaging can be used to look for myelin damage in the brain and spinal cord. Electrical tests of nerves can examine how well neurons send electrical signals. Sometimes a lumbar puncture ("spinal tap") will be performed to sample cerebrospinal fluid for the presence of inflammatory cells.

Steroids, immunosuppressants, and interferon drugs can help treat some symptoms of multiple sclerosis, but so far there is no cure for the disease. Actresses Annette Funicello (1942–2013), Teri Garr (1944–), Selma Blair (1972–), and Jamie-Lynn Sigler (1981–); politician Barbara Jordan (1936–1996); media personality Jack Osbourne (1985–); and talk show host Montel Williams (1956–) were all diagnosed with multiple sclerosis.

See also Action Potential; Cerebrospinal Fluid; Magnetic Resonance Imaging; Myelin

Mushrooms, Hallucinogenic
Type of fungi containing psychoactive compounds capable of distorting the perception of sight, sound, taste,

smell, or touch. Although some mushrooms taste great on a pizza or added to spaghetti sauce, other mushrooms contain chemicals that can cause hallucinations and send people on "trips" within their own minds.

Hallucinogenic mushrooms have been used for religious and recreational purposes by people around the world for centuries. Psilocybin (psilocin) mushrooms and the *Amanita muscaria* mushroom are two primary types of hallucinogenic mushrooms. Mushrooms belonging to the genera *Psilocybe, Stropharia, Conocybe,* and *Panaeolus* may contain the chemicals psilocybin and psilocin. Mushrooms with these chemicals can cause visual and auditory hallucinations starting 30 to

60 minutes after the mushroom is consumed and lasting about 4 hours. Psilocybin and psilocin have a chemical structure that is similar to the neurotransmitter serotonin. In the brain, these chemicals act on serotonin receptors and also reduce the reuptake of serotonin by neurons. Therefore, psilocybin and psilocin boost the activity of neurons that use the serotonin neurotransmitter system.

The *Amanita muscaria* mushroom, also known as "fly agaric" for its ability to attract and kill flies, contains muscimol and ibotenic acid. This mushroom can cause hallucinations and feelings of euphoria 30 to 90 minutes after it is consumed with the intense behavioral effects lasting 2 to 3 hours. Muscimol acts on the GABA neurotransmitter system by exciting GABA receptors on neurons. Because the GABA neurotransmitter system is inhibitory in nature, muscimol acts to reduce the activity of neurons in the brain. Ibotenic acid works on glutamate neuronal receptors and can also be converted into muscimol.

Amateur mycologists searching for hallucinogenic mushrooms must be careful what they pick because their prized fungi may look similar to toxic, deadly mushrooms. In the United States, mushrooms containing psilocybin or psilocin are considered Schedule I drugs,[72] making the possession, sale, transportation, and cultivation of these mushrooms illegal. However, some states have decriminalized the possession of these mushrooms. *Amanita muscaria* mushrooms are legal in the United States, except in the state of Louisiana.

See also GABA; Lysergic Acid Diethylamide (LSD); Neurotransmitters; Serotonin

Myelin
Fatty substance that surrounds the axon of a neuron.
Nerve cells send information down axons using electrical signals (action potentials). These signals travel faster
when an axon is insulated with myelin. Myelin wraps
around an axon to help electrical current within an axon
flow faster. This works in a similar way as wrapping tape
around a leaky water hose to help water flow faster within
the hose.

Myelin is created by glial support cells, called oligo-
dendrocytes, in the central nervous system (brain and
spinal cord) and by Schwann cells in the peripheral nervous system. Myelin covers sections of an axon with a
small space of 0.2 to 2 mm between each segment. These
breaks are called nodes of Ranvier. An action potential
"jumps" from node to node in a process called saltatory
conduction. Saltatory conduction allows an action potential to travel faster through axons with myelin compared to the speed of an action potential in an axon
without myelin.

Damage to the myelin sheath can be caused by genetic disorders, inflammation, and viruses. Such damage may result in sensory, movement, and cognitive
problems associated with neurological disorders, such
as multiple sclerosis, Guillain-Barré syndrome, and peripheral neuropathy.

See also Axon; Glia; Multiple Sclerosis; Neuron; Saltatory Conduction

 arcolepsy
Neurological disorder that disrupts both the
onset of sleep and the sleep cycle. Comedian

and late night TV host Jimmy Kimmel (1967–) has trouble staying awake during the day.[73] He has been known to fall asleep during meetings, while driving, and even in the middle of a TV show he was hosting. Like other people with narcolepsy, Kimmel fights a powerful urge to sleep throughout the day. Some people, but not Kimmel, have narcolepsy with catalepsy, a sudden loss of muscle tone that causes them to fall down. Other people with narcolepsy have a disturbing condition called sleep paralysis, where they are not able to move their limbs immediately after they wake up or just before they fall asleep. Visual, auditory, or tactile hallucinations may also occur just before people with narcolepsy fall asleep.

Narcolepsy appears to occur when rapid eye movement (REM) sleep intrudes while a person is awake. Although the cause of narcolepsy is unknown, there appears to be a genetic factor because the disorder can run in families. Having a genetic predisposition for narcolepsy may make someone more likely to develop the

disorder if they are exposed to a certain environmental trigger, such as an infection. Narcolepsy with catalepsy has been linked to a lack of hypocretin in the brain. Hypocretin is a neuropeptide produced in the hypothalamus that is involved with sleep, arousal, and appetite.

Medications such as central nervous system stimulants and antidepressants can control some symptoms of narcolepsy. Provigil (modafinil) is an FDA-approved stimulant that Jimmy Kimmel and other people use to treat their sleepiness caused by narcolepsy. Good sleep hygiene, including short naps, maintaining a regular sleep schedule, and reducing the amount of alcohol and caffeine before bedtime, can also benefit people with narcolepsy.

See also Neurotransmitters; Sleep

Nerve Agent

Chemical that interrupts the normal function of the nervous system; can be used as a weapon. Originally developed as insecticides, nerve agents have found their way into the arsenals of official government militaries as well as terrorist organizations.

Common nerve agents, such as sarin, soman, VX, and tabun, are classified as organophosphate chemicals. These chemicals have little or no odor and no color. When dispersed in bombs, missiles, or sprays, nerve agents can be absorbed through the skin or inhaled in contaminated air. Vaporized nerve agents can cause symptoms within seconds after people are exposed to the chemicals.

In the body, nerve agents alter the function of the neurotransmitter acetylcholine. When molecules of acetylcholine are released from axon terminals, they move across the synaptic gap and bind with receptors on other neurons, organs, and muscles where they act. The action of acetylcholine is stopped by an enzyme called acetylcholinesterase that breaks apart the acetylcholine molecule. Nerve agents inactivate acetylcholinesterase, and therefore acetylcholine works unchecked. An overactive acetylcholine system results in a variety of symptoms, including paralysis, nausea, breathing problems, vomiting, drooling, coma, and death.

People exposed to nerve agents can be treated with drugs such as atropine and pralidoxime chloride. Atropine blocks acetylcholine receptors and pralidoxime blocks the nerve agent from acting on acetylcholinesterase. The result of either drug is to normalize acetylcholine transmission.

Although the production, stockpiling, and use of nerve agents has been banned by many countries through the Chemical Weapons Convention,[74] tons of these toxic substances still exist, waiting for disposal.

Neuroethics

Field of study that focuses on ethical issues raised by brain research. Neuroethics seeks to understand how advances in our understanding of the brain impact individuals and society. Although brain research has led

to discoveries that have improved the lives of millions of
people through the development of drugs and therapies
for people with psychiatric and neurological disorders,
important questions remain about how brain research
should be conducted and how experimental data should
be used.

People from all walks of life should be interested in
neuroethics because advances in neuroscience may create potential problems that affect their lives. For example,
scientists and engineers are currently building novel neurotechnologies, such as brain-computer interfaces and
deep brain stimulators, that connect the nervous system
with machines. The safety and possible side effects of
these devices are not always known. In the United States,

neurotechnologies intended for medical use must abide by US Food and Drug Administration (FDA) regulations and demonstrate that they are safe and effective. However, many direct-to-consumer neurotechnologies—for example, consumer electroencephalography devices and transcranial direct current stimulation units—are not classified as medical devices and therefore fall outside of US FDA oversight.[75]

The security of neurotechnologies must be protected to avoid unintentional disruption of these devices or purposeful malicious alteration of their functions. Furthermore, data resulting from neuroscientific research must be kept private to protect an individual's identity, the involuntary release of personal recordings, and the unapproved release of information to others for financial gain.

Other neuroethical issues focus on responsibility, justice, and changes in identity. Some devices or drugs created by neuroscientists and neuroengineers may have a limited period of usefulness. If these innovations fail, it is unclear who will be responsible for the consequences or their maintenance. Scientific research is also very expensive. Therefore, it is unclear who would make decisions regarding funding of scientific areas and who would have access to the resulting products. If the therapy or treatment is expensive, this may divide society into groups of people who can afford and benefit from the product and those who cannot. Finally, neuroscientific research has the potential to change a person's personal identity. Neurotechnologies and neuroactive drugs can alter memory, mood, and personality. This may change how people view themselves and how others view people who use these new devices or drugs.

Brain researchers will likely continue to develop new technologies to interact and alter the function of the nervous system, but whether and how such discoveries should be used to benefit society remain to be debated.

See also Brain Initiatives

Neuron

Specialized cell in the nervous system that transmits information. Neurons (nerve cells) use electrochemical messages to communicate with other neurons, muscles, organs, and glands. The adult human brain contains about eighty-six billion neurons,[76] and there are billions more neurons in the spinal cord and peripheral nervous system.

Neurons have a variety of shapes and sizes, but all have the same basic parts. Each neuron has dendrites that receive information from other neurons and bring electrical signals to a neuron's cell body. The smallest neurons have cell bodies with diameters of 4 microns while large neurons can have a cell body of 100 microns in diameter. A neuron's cell body contains many of the same organelles as other cells, such as a nucleus, nucleolus, Nissl bodies, endoplasmic reticulum, ribosomes, Golgi apparatus, and mitochondria. Neurons also use microfilaments and neurotubules for structural support and to transport materials within a neuron. The cell body is connected to an axon that takes information away from the cell body. An axon can be very long: for example, a neuron in the spinal cord can have an axon stretching 1 meter all the way to a muscle in the foot.

Messages can be sent quickly down an axon with an electrical signal called the action potential. Some axons

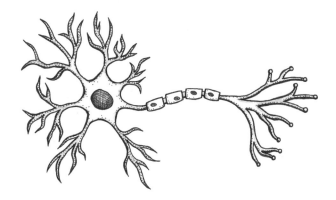

are surrounded by a myelin sheath that increases the speed of an action potential. When an action potential reaches the terminal at the end of an axon, chemical neurotransmitters are released into a gap (synapse) that increases or decreases the activity of another cell. The flow of information is one way—from dendrites, to cell body, to axon, to axon terminal.

Neurons can be classified by the number of dendrites and axons that extend from the cell body. Bipolar neurons have only one dendrite and one axon extending from the cell body. Pseudounipolar neurons have a single process that extends from the cell body that then splits, with one side going toward the spinal cord and the other side going to skin or muscle. Multipolar neurons have multiple dendrites attached to the cell body, but they still have only one axon.

Another way to classify neurons is by the direction that they send information. Sensory neurons send information from the skin, eyes, nose, tongue, and ears into the brain or spinal cord. Motor neurons send information out of the brain and spinal cord to muscles or glands. Interneurons are responsible for communication between neurons in a brain area and between sensory neurons and motor neurons.

Neurons are the oldest cells in the body: many of the neurons people are born with stay with them for their entire lifetime.

See also Action Potential; Axon; Cell Body; Dendrite; Glia; Myelin; Neurotransmitters; Synapse

Neuroplasticity

Ability of the nervous system to reorganize neural pathways. The brain undergoes changes throughout a person's lifetime in response to genetic and environmental factors. These changes help people learn and memorize new information and can repair damage to neural circuitry.

Neuroplasticity is especially evident during normal brain development when the immature brain begins to process sensory information. Genetic coding provides the blueprint for neurons to find their proper connections from sensory receptors to particular brain areas. Thousands of synaptic connections are made during this process. In fact, many more connections are made during the early years of brain development than are necessary. As an organism interacts with its environment, synaptic connections that are used will be strengthened and those that are not used will be pruned away and lost. Synaptic pruning reshapes and refines neural pathways, and neurons that are not used will die. This mechanism allows the brain to adapt to its environment. Exposure to drugs, injury, or disease may interfere with this normal developmental process.

Our abilities to learn and form new memories also involve neuroplastic changes by altering synaptic connectivity between neurons.[77] Repetition and reinforcement of behavior leads to structural and biochemical changes within synapses that strengthen connections between neurons. Rewiring of brain pathways also occurs after the brain has been damaged. For example, after a stroke, the brain can be rewired such that the function of damaged areas can be taken over by neurons in neighboring brain regions.

A better understanding of the factors involved with neuroplasticity holds the promise of novel therapies for traumatic brain injury and neurodegenerative diseases. These treatments would manipulate the way the brain rewires itself to create new pathways to repair damage.

See also Brain, Development; Stroke; Synapse

Neurotoxin

Synthetic or natural substance that can disrupt the structure and function of the nervous system. Neurotoxins include a wide variety of chemicals, including some insecticides and pesticides. In nature, neurotoxins are found in the venoms of some insects, mollusks, scorpions, spiders, and snakes, as well as in chemicals found in some fish, amphibians, and plants. Some chemical elements, notably lead and mercury, are also classified as neurotoxins.

Different neurotoxins act on neurons in different ways. Some neurotoxins block receptors for specific neurotransmitters while other neurotoxins stimulate the

release of neurotransmitters. Many neurotoxins block the opening of neuronal ion channels, thereby interfering with the generation of action potentials.

Although it is a good idea to avoid neurotoxins, some people intentionally expose themselves to these substances. For example, the pufferfish (fugu) is considered to be a delicacy served in Japanese restaurants. Fugu contains tetrodotoxin (TTX), a neurotoxin that poisons sodium channels and prevents the generation of action potentials. Consuming a small amount of TTX in fugu may cause a slight tingling or numbness around a diner's lips and mouth, but larger amounts can be fatal. Botulinium toxin, found in the product Botox, is another neurotoxin that people purposely pursue. Botox injected into muscles blocks the release of the neurotransmitter acetylcholine. This action works to smooth wrinkles and relieve muscular spasms.

See also Action Potential; Lead; Loewi, Otto; Mercury; Nerve Agent

Neurotransmitters

Chemical messengers that transmit information across the synapse to communicate from one neuron to another neuron, muscle, or gland. The 3-pound adult human brain contains more than 200 different chemicals that enable neurons to communicate with other cells. The brain might be considered a soup containing these chemical messengers as the special ingredients responsible for perceptions, actions, and emotions.

Some neurotransmitters, such as acetylcholine, dopamine, and serotonin, are synthesized within a neuron's synaptic terminal while others, such as neuroactive peptides, are made in the cell body of a neuron and then transported to the synaptic terminal. Regardless of where they are made, chemical neurotransmitters are stored in small sacs (vesicles) in the synaptic terminal. When an electrical impulse (action potential) travels down an axon to a neuron's synaptic terminal, its triggers vesicles containing neurotransmitters to move toward the membrane. The vesicle membrane then fuses with the membrane of the axon terminal. This results in the release of the neurotransmitters into the synaptic cleft.

Molecules of neurotransmitters make their way across the synapse cleft, where they may attach to receptor sites on the other side of the synapse. Like a lock and key, a neurotransmitter must have the proper shape to fit into a receptor site. Depending on how a receptor site responds to a neurotransmitter, the receiving cell will become more (excited) or less (inhibited) likely to fire its own action potential. A receiving cell may have thousands of synapses and must add all of the excitatory and inhibitory events. If the amount of excitation in the receiving cell exceeds a certain level, it will fire an action potential.

The action of a neurotransmitter can be stopped in several ways. First, neurotransmitters may simply move away from a receptor site so they cannot act. Second, enzymes can change the physical structure of a neurotransmitter molecule so that a neurotransmitter no longer fits into the receptor site. Third, glial cells—specifically, astrocytes—can remove neurotransmitters from the synaptic cleft. Lastly, in a process called reuptake, neurotransmitter molecules can be removed from the synaptic cleft and put back into the axon terminal that released them.

Nitric oxide and carbon monoxide are two gases that act as neurotransmitters. These gases are not stored in synaptic vesicles but instead diffuse out of the neuron. The gases can then enter another cell, where they modulate neuronal activity.

See also Action Potential; Dopamine; GABA; Glia; Neuron; Neurotoxin; Serotonin; Synapse

Nicotine

Central and peripheral nervous system stimulant found in plants such as tobacco. Nicotine works by binding to receptors on neurons that use the neurotransmitter acetylcholine. More specifically, nicotine activates nic-

otinic acetylcholine receptors. The activation of these receptors in various areas of the brain results in the release of other neurotransmitters such as dopamine, norepinephrine, serotonin, and endorphins (endogenous opioids). Nicotine also acts on nicotinic receptors in the peripheral nervous system. The combined central and peripheral effects of moderate doses of nicotine are constriction of peripheral blood vessels, increased heart rate, higher blood pressure, and increased attention.

Nicotine also activates reward pathways in the brain that provide feelings of pleasure. Repeated stimulation of the reward pathway by nicotine changes the sensitivity of neurons to neurotransmitters such as dopamine. This may result in brain changes responsible for addiction to nicotine and underlie withdrawal symptoms when people do not have access to nicotine.

The tobacco plant is called *Nicotiana tabacum*, named after French diplomat Jean Nicot (1530–1600), who was

the first person to import tobacco to France. The to-
mato, potato, eggplant, and green pepper plants contain
nicotine in small quantities.

See also Dopamine; Neurotransmitters; Serotonin

Nootropics

Substances intended to improve brain functions such
as memory, problem solving, attention, creativity, and
motivation. Most people would like to have a better
memory and be able to learn faster. Nootropics, also
known as smart drugs or cognitive enhancers, are pur-
ported to provide people with a mental tune-up that
improves these and other mental abilities.

Nootropics seek to enhance cognition by increasing
brain metabolism, increasing brain blood flow, changing neurotransmitter activity, or protecting the brain
from damage. Some chemicals may affect the brain in
these ways, but research investigating the beneficial effects of these products on mental function is not con-
clusive.[78] Nevertheless, companies producing sports
drinks, power bars, and dietary supplements have all
jumped into the smart drug business by adding herbs or
chemicals such as phenylalanine, choline, taurine, and
ginseng to their products. Although these companies
may have great sales pitches, there is little evidence
to support the claims that their products provide any
cognitive benefits to healthy people.

See also Neurotransmitters

ccipital Lobe
Part of each cerebral hemisphere located behind the parietal lobe and temporal lobe and

above the cerebellum; responsible for processing information about vision. The occipital lobe (visual cortex) contains several areas that process information from the eyes. Each of these areas is responsible for decoding different aspects of a visual image. The primary visual cortex (VI or area 17) receives information from the lateral geniculate nucleus of the thalamus. VI on the left side of the brain receives information about the right visual field and VI on the right side of the brain receives information about the left visual field. The primary visual cortex is often called the striate cortex because of its striped appearance. Neurons in VI respond to the orientation and direction of motion. For example, VI neurons are tuned to respond best to lines of a specific angle or movement in a particular direction.

Information processed in VI is sent on to other areas of the occipital lobe (V2, V3, V4) and temporal lobe, where features about color, shape, and other more complex characteristics are extracted. Working together, all areas in the occipital lobe contribute to the final perception of what we see.

David Hubel (1926–2013) and Torsten Wiesel (1924–) won the 1981 Nobel Prize for Physiology or Medicine for their discoveries about how information is processed in the visual system.

See also Temporal Lobe

Optogenetics

One of the newest techniques in a neuroscientist's toolbox that uses light and genetics to control the activity of nerve cells. The keys to the success of optogenetics are opsins, proteins that open ion channels when exposed to

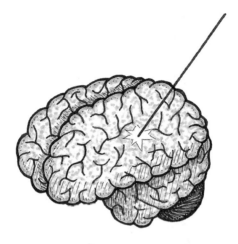

light.[79] Different opsins are sensitive to different wave-lengths of light and allow different ions to move across cell membranes. Scientists reasoned that opsins could be used to take advantage of how neurons send electrical messages by the exchange of ions between the inside and outside of a neuron.

In the early 2000s, researchers genetically modified a virus with a light-sensitive ion channel gene. When injected into the brain, the virus could get its DNA into the DNA of a neuron. Neurons with this new DNA would then express the light-sensitive ion channels on their membranes. Therefore, when light delivered via a surgically implanted fiber-optic device strikes these neurons, light-sensitive ion channels open and ions can flow across the membrane. For example, if light causes a sodium ion channel to open, sodium ions will flow into a neuron, causing a depolarization and an action potential. Other opsins can be used so that light opens channels to other ions (e.g., chloride ions) to inhibit the firing of action potentials. Viruses can be tailored for specific types of channels and neurons and then injected into specific places in the brain.

Optogenetics has the potential to be more than a lab tool; the technique could revolutionize the treatment of neurological disease. For example, someday targeted light that activates specific neurons may improve the memory of people with Alzheimer's disease or reduce movement problems in people with Parkinson's disease.

See also Action Potential; Alzheimer's Disease; Parkinson's Disease

Ossicles

Set of three small bones in the middle ear. The stapes (stirrup), malleus (hammer), and incus (anvil), collectively known as the ossicles, help transmit sound pressure waves from the outer ear to the inner ear. The three bones form a chain of levers that are connected by joints. On one side of the middle ear, the eardrum (tympanic membrane) is attached to the malleus by ligaments. The malleus is connected to one side of the anvil and the stapes is connected to the other side of the anvil. Finally, the stapes is attached to the oval window. When sound pressure waves enter the outer ear, they vibrate the eardrum, which causes the ossicles to move. When the stapes moves, it vibrates the membrane of the oval window, which causes fluid movement in the cochlea. This arrangement of membranes and levers transfers sound waves to the inner ear, where the waves are converted into electrical nerve impulses for the brain.

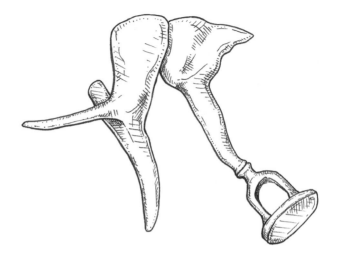

The stapes is the smallest bone in the body, about 2.8 mm in length and 3.3 mm in height[80]—good information for the next time anyone is on a game show.

See also Cochlea

Parietal Lobe

Part of each cerebral hemisphere located behind the frontal lobe, on top of the temporal lobe, and in front of the occipital lobe; responsible for processing sensory information from the skin and for the perception of the body in space. The parietal lobe contains the somatosensory cortex with neurons that respond to touch, pressure, temperature, and pain on the skin. The somatosensory cortex is organized such that more brain tissue is devoted to areas with higher sensitivity. For example, the fingers, hands, mouth, and face all have larger areas of somatosensory cortex than the toes, feet, legs, and back.

Parkinson's Disease

Progressive neurodegenerative disorder caused by damage to the dopamine system of the brain. In 1817, James Parkinson (1755–1824) described a disorder he called the "shaking palsy," where patients had involuntary tremors in the hands, muscle weakness, abnormal posture, and slow movements.[81] It took more than a century after this description to discover the cause of this condition.

People with Parkinson's disease show four cardinal symptoms: (1) tremor or shaking of the extremities, (2) rigidity and muscle stiffness, (3) slow movement, and (4) impaired balance and posture. In addition to having problems with movement, people with Parkinson's disease may experience nonmotor symptoms, such as pain, difficulty swallowing, depression, and sleep problems.

Although the exact cause of Parkinson's disease is not known, a combination of genetic and environmental factors is likely responsible for the condition. The changes within the brains of people with Parkinson's disease are well known: neurons that produce the neurotransmitter dopamine are damaged and die. When enough dopamine-producing neurons die, a person will exhibit signs of the disease.

There is no cure for Parkinson's disease, so treatment of the disorder focuses on reducing symptoms that interfere with a person's daily activities. The simple administration of dopamine is not an effective treatment because this neurotransmitter does not cross the blood-brain barrier. However, levodopa, the precursor to dopamine, does cross into the brain, where it is converted into dopamine. Although levodopa does not stop the progression of Parkinson's disease, it is usually effective in minimizing many of the disorder's symptoms. Other medications that mimic the effects of dopamine or slow the breakdown of dopamine in the brain can

MUHAMMAD ALI

also be used to treat the disease. Some people with Parkinson's disease, such as actor Michael J. Fox (1961–), elect to undergo brain surgery to remove a small part of the thalamus or globus pallidus to reduce tremors. Other people with Parkinson's disease benefit from deep brain stimulation, where an electrode is inserted into the brain to deliver electrical signals to control abnormal movements.

Celebrities who have been diagnosed with Parkinson's disease include boxer Muhammad Ali (1942–2016); former president George H. W. Bush (1924–2018); comedian Billy Connolly (1942–); singers Neil Diamond (1941–), Ozzy Osbourne (1948–), and Linda Ronstadt (1946–); actors Michael J. Fox (1961–), Alan Alda (1936–), and Bob Hoskins (1942–2014); Rev. Jesse Jackson (1941–); film critic Leonard Maltin (1950–); Pope John Paul II (1920–2005); and former attorney general Janet Reno (1938–2016).

See also Blood-Brain Barrier; Dopamine

Penfield, Wilder (1891–1976)

American-Canadian neurosurgeon who developed several surgical procedures to treat patients with epilepsy and other neurological disorders. When he attended college at Princeton University, Penfield excelled in the classroom as a student and on the field as a football player. With a Rhodes Scholarship, Penfield began his studies at Oxford University in 1915 under the mentorship of Charles Sherrington (1857–1952).[82] After receiving his bachelor's degree in physiology from Oxford, Penfield returned to the US and received his medical degree from Johns Hopkins University in 1918. Penfield refined his skills as a neurosurgeon at New York Presbyterian Hospital and then at McGill University in Montreal, Canada, where he established the Montreal Neurological Institute in 1934.

With help from fellow clinicians, Penfield developed the Montreal procedure—a method that allowed surgeons to perform brain operations while their patients were still awake. During this surgery, local anesthetics are used to numb the scalp while the brain is exposed. Touching the brain does not cause any sensation or pain because the brain does not contain any receptors for pain. Penfield stimulated the brain with electricity to cause sensations or bodily movements in his patients. Therefore, patients could describe the resulting sensations and doctors could see movement caused by stimulating different parts of the brain. The technique helped surgeons identify critical brain areas and avoid damaging those areas while removing tumors or tissues that produced seizures.

Penfield is revered as one of the greatest neurosurgeons of all time and was even honored on a stamp issued by Canada in 1991.

See also Epilepsy; Sherrington, Charles Scott

Phrenology

Method that uses the features of a person's skull to reveal personality traits and mental abilities. German physician Franz Joseph Gall (1758–1828) was convinced that specific areas of the brain were responsible for particular behaviors and cognitive functions. Therefore, he reasoned that the shape of the skull should mirror the shape of the brain, with more prominent bumps on the skull reflect-

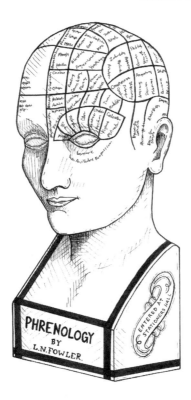

ing more brain area devoted to a particular behavioral characteristic. Gall named his system cranioscopy, from the Greek words meaning "skull" and "vision." Later proponents changed the name to phrenology, meaning "mind" and "study."

Gall's idea was spread by others, such as Johann Kaspar Spurzheim (1776–1832) and George Combe (1788–1858). Gall assigned behaviors and faculties to locations on the skull without knowledge of the function of brain tissue below. To create his system, Gall used the heads of his friends he thought shared common mental characteristics to define areas. The skulls of criminals in jail or patients in mental hospitals helped refine the method by assigning aberrant behaviors to places on the skull. Based on these observations, Gall attempted to correlate particular mental faculties to bumps and depressions on the surface of the skull. Gall assigned twenty-six different characteristics to places on the skull while Spurzheim and Combe assigned additional traits, such as combativeness, secretiveness, hope, and wonder.

Although phrenology was met initially with suspicion by scientists and religious leaders, it did enjoy a good deal of public interest. Nevertheless, advances in our knowledge about the brain have completely discredited phrenology and have placed phrenology on the pile of pseudoscience.

There is a kernel of scientific fact buried within the fiction of phrenology. Gall's idea of functional localization—theorizing that specific areas of the brain are involved with particular functions—is correct. Phrenology has also given us phrases we still use today: "highbrow," "lowbrow," and "getting your head examined" all have their origins in phrenology.

Positron Emission Tomography (PET)

Functional brain imaging method. PET imaging is used to study brain activity to help diagnose disease and study normal neurological functions. To obtain a PET image, a radioactive material (e.g., radioactively labeled glucose, oxygen, fluorine, or carbon) must first be swallowed, injected, or inhaled by the person undergoing the procedure. The radioactive material travels through the bloodstream into the brain, where it accumulates in areas that use it. For example, oxygen and glucose accumulate in brain areas that are metabolically active. Therefore, brain areas that have higher activity consume more material and are more radioactive. Radioactive chemicals that attach to specific neurotransmitter receptors can be used to study psychiatric disorders. In this case, brain areas with a higher density of specific neurotransmitter receptors have more radioactive label.

When the radioactive material breaks down, it gives off a neutron and a positron. When a positron hits an electron, both are destroyed and two gamma rays are released. Gamma ray detectors around a person's head record the brain area where the gamma rays are emitted. The location of the gamma rays can be reconstructed to get a picture of the activity.

Unlike computed tomography and magnetic resonance imaging methods that show only brain structure, PET provides a record of brain function. Therefore, PET can be used to investigate which parts of the brain are active when people perform specific tasks. However, PET is rather expensive and requires special laboratories to prepare the radioactive materials. Some people may be alarmed at the prospect of having radioactive material injected into their bloodstream. However, the dose of radiation is kept as low as possible to minimize the risks of the procedure but still provide the necessary information to create a useful image.

See also Computed Tomography; Magnetic Resonance Imaging; Neurotransmitters

Prosopagnosia (Face Blindness)

Neurological condition characterized by the inability to recognize faces. Imagine an artist who paints portraits but who is unable to recognize faces, even familiar ones. This is how painter Chuck Close (1940–2021) went about his art: he sold masterpieces for millions of dollars, but could not recognize the people that he painted by looking at their face.

Prosopagnosia is much more than simply forgetting the name of someone you meet. People with prosopagnosia have trouble recognizing the faces of familiar people, and they are unable to distinguish the faces of friends and family members from those of strangers. Some people with prosopagnosia cannot even recognize their own face in a mirror or photograph.

Prosopagnosia can be caused by damage (e.g., stroke, trauma, infection) to the fusiform gyrus, an area in the temporal lobe of the brain.[83] The fusiform gyrus is part of a neural circuit responsible for facial perception and memory. Some people with prosopagnosia are born with the condition.

To cope with the inability to recognize faces, people with prosopagnosia can learn to identify other people by their clothing, hairstyle, voice, or walking patterns. Of course, these strategies do not always work if people change their clothes or get a haircut.

See also Stroke; Temporal Lobe

Rabies

Brain disease caused by a virus that infects the central nervous system. Hearing someone yell "rabid dog" usually sets off a wave of fear and panic.

Rabid dogs often first appear tired, but then they may develop a fever, vomit, drool ("foam at the mouth"), and avoid light. Infected dogs may show other strange, unpredictable, aggressive behavior and fear of water (hydrophobia) is possible too. People infected by rabies show similar behaviors, and after these symptoms appear the result is almost always death. The only thing that can be done for these patients is to keep them as comfortable and free of pain as possible.

The rabies virus is transmitted through the bite or scratch from an infected animal. Although dogs cause most of the rabies infections in people, bats, raccoons, foxes, skunks, and any other mammal can harbor the rabies virus. Once inside the body, the virus moves through nerves up to the spinal cord and brain. In the brain, the virus causes inflammation and swelling (encephalitis).

Vaccines can be highly effective in preventing rabies in dogs. In fact, in the US, a successful animal vaccination program has virtually eliminated human rabies. The Centers for Disease Control and Prevention reports only one to three cases of human rabies each year in the US. Unfortunately, rabies is more common in other parts of the world and causes an estimated 59,000 deaths each year.[84]

Ramón y Cajal, Santiago (1852–1934)

Spanish neuroanatomist who is often called the father of modern neuroscience for his foundational work on the structure of the neuron. Who would have thought that the father of modern neuroscience would be such a rebel? As a youth in a small village in Spain, Ramón y Cajal was not the best student and he attended several different schools. When he was eleven years old, Cajal spent three days in jail for destroying a neighbor's gate with a homemade cannon.[85] After graduating from medical school in 1873, Cajal served as a doctor in the Spanish Army stationed in Cuba. While in Cuba, he contracted malaria and dysentery. After his return to Spain, Cajal studied at the University of Madrid, where he received his PhD.

Cajal turned his attention to the nervous system when he learned of Camillo Golgi's "black reaction," a technique that completely stains a random subset of neurons so researchers can see the entire structure of a neuron with a microscope. With improvements to Golgi's method, Cajal was able to provide detailed descriptions of neural structures. His careful observations led him to theorize that neurons were not linked physically but instead were separated from one another. Golgi disagreed with this theory and proposed that neurons were connected in a netlike fashion.[86] Although Cajal was later proved to be correct, both he and Golgi shared the 1906 Nobel Prize in Physiology or Medicine for their work.

Cajal was certainly productive in his profession and personal life: he published more than one hundred scientific papers, received a Nobel Prize, and was an accomplished artist, photographer, and writer.[87, 88] He also produced seven children with his wife, Silveria Fañanás García (1854–1930).

SANTIAGO RAMÓN Y CAJAL

Reflex

Automatic movement that does not require conscious control. Reflexes are fast, fixed, automatic movements made in response to stimulation of the senses. The primary job of reflexes is to protect the body from injury. For example, the withdrawal reflex protects the skin from damaging pressure or temperature, the pupillary reflex prevents damage to the retina caused by bright light and the knee-jerk reflex helps maintain balance and posture.

Reflexes involve circuits of neurons, starting with the activation of sensory receptors. Sensory receptors then send messages to neurons in the spinal cord or brain. These spinal cord or brain neurons send signals back to the body to respond to the original sensory event. The entire response does not require any conscious thought.

Most people have had their "knee-jerk" reflex (patellar reflex) tested during a visit to their doctor. During this test, the patellar tendon is tapped with a small rubber hammer. The tap stretches sensory receptors called muscle spindles in the quadriceps muscle. The muscle spindles send electrical signals to the spinal cord where they synapse on motor neurons. The motor neurons respond by sending messages to thigh muscles to contract and kick out the leg. Other signals from motor neurons are responsible for relaxing leg muscles that work to oppose the kick. Abnormal reflexes—for example, slow, absent, or excessive responses—may indicate problems within the neuronal circuit or muscle.

Restless Legs Syndrome (RLS)

Neurological disorder characterized by an urge to move the legs usually when a person is sitting or lying down.

Imagine trying to go to sleep, but feeling as if your legs were burning or had bugs crawling on them. Moreover, the only way to get rid of those sensations is to move your legs. These are the symptoms of restless legs syndrome, a condition that affects 5%–15% of the population. Because people with RLS have trouble falling asleep and staying asleep, they often are sleepy during the day.

Although the exact cause of RLS is not known, a genetic component is likely because the disorder runs in families. Environmental factors also likely contribute to RLS because the condition occurs more often in people with low levels of iron, kidney disease, diabetes, Parkinson's disease, nerve disorders, and spinal cord injuries. Medications such as tricyclic antidepressants, antipsychotics, and caffeine can worsen RLS symptoms.

Drugs used to treat Parkinson's disease are sometimes effective in relieving RLS symptoms. For example, levodopa and other medications that boost brain levels of the neurotransmitter dopamine can reduce tremor and other signs of Parkinson's disease and can also reduce RLS symptoms. Antianxiety drugs, opiates, and anticonvulsants may also improve sleep, reduce unpleasant sensations, and reduce leg movements in people with RLS. Some people with RLS can benefit from moderate exercise, hot baths, massage, and regular sleep routines.

See also Dopamine; Neurotransmitters; Parkinson's disease

Retina

Innermost layer of the eye composed of five cell layers: ganglion cells, amacrine cells, bipolar cells, horizontal cells, and photoreceptors. The lens and cornea of the

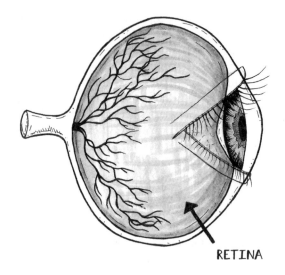

RETINA

eye focus light rays on the retina to excite two types of light-sensitive photoreceptors called rods and cones. Rods and cones have molecules (photopigments) that absorb certain wavelengths of light. Light changes the shape of these photopigments to cause an electrical signal to be sent out of the retina to the brain. Signals from both rods and cones are sent to the brain, where a picture of the external world is created. The human retina contains about 120 million rods and 6 million cones.

Rods contain only one type of light-sensitive pigment and are most sensitive to changes in light intensity, shape, and movement. Cones are not as sensitive to light as rods and require bright light to work. However, because cones are most sensitive to one of three different colors (green, red, or blue), they are used for color vision. The fovea is the central region of the retina that provides for the clearest vision because it has only cones and no rods. People who lack one or more types of cone photoreceptors have difficulty distinguishing some colors from others. In other words, they are color-blind. The incidence of color blindness is about 8% in males and 0.5% in females.

Photoreceptors in the human retina are sensitive to wavelengths within the visible spectrum of light (between 380 and 760 nanometers). Some birds, fish, and butterflies have photoreceptors that can detect light in the ultraviolet range of light (between 300 and 380 nanometers).

See also Blind Spot

Saltatory Conduction

Movement of an action potential from node to node along a neuron's axon. Glial cells form myelin that wraps around and insulates axons. Each myelin sheath covers only a section of an axon, leaving small gaps between each insulated section. These gaps (0.2–2 mm in width) are called nodes of Ranvier. In an axon wrapped with myelin, the nodes of Ranvier are the only places where sodium and potassium ions can flow across the membrane to generate an action potential. At each node of Ranvier, the action potential is regenerated, and this electrical signal jumps from node to node in rapid succession as it travels down an axon. This process is called saltatory conduction, which comes from the Latin phrase meaning "to leap."

Saltatory conduction provides a fast way for action potentials to travel down an axon. Large-diameter myelinated axons can send action potentials at speeds up to 120 m/s (432 km/hr; 268 miles/hr), while the fastest action potentials in unmyelinated axons are only about 2 m/s (7.2 km/hr; 4.5 miles/hr).

See also Action Potential; Axon; Glia; Myelin

Schizophrenia

Mental illness characterized by disturbed patterns of thinking, emotions, and movement. The word *schizophrenia* comes from the Greek words meaning "split" and "mind" to describe how people with schizophrenia have made a split from reality.

People with schizophrenia have delusions, for example, they believe that other people can control their behavior. Hallucinations such as hearing voices, seeing lights or other objects, smelling odors, or feeling sensations on their skin are also common symptoms of schizophrenia. Some people with schizophrenia have speech problems or they may have abnormal movement. The absence of normal behavior, such as withdrawal from social interactions, lack of emotion, and reduced motivation can also be signs of schizophrenia.

Multiple factors increase the risk that someone will develop schizophrenia. Genetics likely plays a significant role in schizophrenia, and the disorder does run in families. Evidence for a genetic link is also found in twin studies that show a higher incidence of schizophrenia in identical twins compared to the incidence of the disorder in fraternal twins. A person's family environment, social interactions, infections, and early trauma may also contribute to the cause of schizophrenia.

Schizophrenia has been linked to changes in brain structure and function and neurotransmitter levels. On average, people with schizophrenia have larger brain ventricles than people without schizophrenia. Reduced thickness of the cerebral cortex and reduced myelination of cortical axons are also features of schizophrenia.[89, 90] Many studies suggest that an overactive dopamine neurotransmitter system in the brain contributes to schizophrenia and drugs that block dopamine reduce schizophrenic symptoms.

Swiss psychiatrist Eugen Bleuler (1857–1939) coined the term *schizophrenia* in 1908.[91]

See also Dopamine; Neurotransmitters; Ventricles

Serotonin

Neurotransmitter involved in a wide variety of functions, including the regulation of mood, memory, pain, sleep, and digestion. Most of the serotonin in the human body is found in cells lining the digestive tract.[92] These cells help regulate digestion and other gastrointestinal processes. In the brain, serotonin is produced by neurons in the midbrain that send projections to multiple areas of the nervous system, such as the cerebral cortex, thalamus, cerebellum, medulla, hypothalamus, and spinal cord.

Drugs that target the serotonin system have proved to be useful in the treatment of psychiatric and neurologic disorders, such as depression, panic disorder, obsessive-compulsive disorder, and anxiety. For example, some antidepressants selectively block the reuptake of serotonin after it is released into a synapse. These selective serotonin reuptake inhibitors (SSRIs) therefore increase the availability of serotonin to bind with receptors and transmit messages to other neurons.

Serotonin is an ingredient contained in neurotoxins found in some scorpions, spiders, insects, snakes, sea urchins, and hornets.[93] The injection of venom containing serotonin produces intense pain and inflammation.

See also Brainstem; Neurotoxin; Neurotransmitters; Twarog, Betty Mack

Sherrington, Charles Scott (1857–1952)

English scientist who pioneered the field of neurophysiology. Sherrington not only transformed the study of the nervous system with rigorous methods and innovative theories, he was a dedicated teacher who trained many scientists and physicians who became giants in the fields of neurosurgery, neurology, and neuroscience.

Sherrington studied medicine at St. Thomas's Hospital with additional courses in physiology at Gonville and Caius College at the University of Cambridge. His first interest was in bacteriology before turning his attention to the nervous system.[94] In 1913, Sherrington took a position at the University of Oxford, where he remained until his retirement in 1936.

The scientific achievements of Sherrington revolutionized our understanding of the nervous system. Sherrington detailed how sensory and motor functions were integrated within the spinal cord, showed that inhibition was an important component in neuronal function, and described how neurons form circuits. Sherrington is credited with coining the term *synapse* in 1897. In 1906, Sherrington published *The Integrative Action of the Nervous System,* a collection of ten Silliman lectures he delivered at Yale University in 1904.

Sherrington received many honors and was knighted by King George V in 1922. He also shared the 1932 Nobel Prize in Physiology or Medicine with Edgar Douglas Adrian (1889–1977) for work on the function of neurons.

See also Reflex; Synapse

Sleep

Altered state of consciousness accompanied by reduced physical activity and diminished response to sensory stimulation. Sleep may look like a time of inactivity, but brains are not idle. Instead, while the body appears at rest, the brain is a hive of activity, moving through regular periods of different states.

The development of electroencephalography (EEG) has allowed researchers to measure the electrical activity of the brain at any time of the day or night. In the early

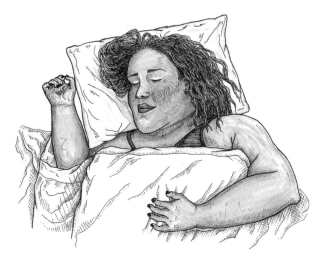

1950s, sleep researchers noticed that the electrical activity of the brain follows a regular, predictable cycle each night. Sleep is classified into two main stages: nonrapid eye movement (NREM) sleep and rapid eye movement (REM) sleep. NREM sleep can be divided into several different stages of sleep characterized by brain waves with different frequencies and sizes. As we transition from wakefulness to sleep, we first enter stage 1 NREM sleep with small-sized brain waves with relatively fast frequencies. Stage 2, stage 3, and stage 4 of NREM sleep follow with larger brain waves with slower frequencies. The EEG rolls back up through the NREM stages until a phase of REM sleep begins. The progression through the different stages of sleep then begins again. Each cycle through NREM and REM sleep stages takes 90 to 120 minutes.

Most dreaming occurs in REM sleep. During REM sleep, a person's eyes move back and forth rapidly. If people awaken during REM sleep, they often say that they were just dreaming. The EEG patterns during REM sleep and wakefulness are very similar with small, high-frequency brain waves. However, during REM sleep, the brain generates signals that prevent voluntary muscles from moving. This safety mechanism stops us from acting out our dreams.

Sleep may help the body recover from the stress and strain of physical and mental activity performed while a person is awake. Sleep may also serve to protect animals by keeping them quiet when threats cannot be seen.

Humans spend about 8 hours or a third of each day asleep. Some bats sleep almost 20 hours each day and giraffes sleep only about 2 hours each day.

See also Electroencephalography

Society for Neuroscience

Professional organization of scientists from around the world who study the nervous system. Founded in 1969 by a group of twenty scientists, the Society for Neuroscience (SfN) has grown to include more than 37,000 members from more than ninety-five countries. In addition to providing professional development opportunities for its members, the SfN is involved with educational and community outreach, science advocacy, and public policy engagement. The SfN also publishes *JNeurosci* (*Journal of Neuroscience*) and *eNeuro*, with papers covering all aspects of neuroscience.

The annual SfN conference attracts approximately 30,000 neuroscientists who present their research in the form of posters, lectures, workshops, and symposiums. This conference is often the first time graduate students present their work to other neuroscientists.

Sperry, Roger Walcott (1913–1994)

American neuroscientist. Although Roger Sperry is best known for his Nobel Prize–winning work with "split-brain" patients, he started his career with pioneering work investigating the developing nervous system. Sperry's early research provided evidence that neurons created connections based on genetic codes and chemical signaling.

Sperry turned his attention to split-brain research in the 1960s.[95] This work involved the study of a unique group of patients who underwent brain surgery to control epileptic seizures. During this operation, neurosurgeons cut the corpus callosum, the large group of axons that connect the right and left cerebral hemispheres. Prior to Sperry's investigation, researchers found that patients with a split corpus callosum had no cognitive changes after the surgery. Not satisfied with those studies, Sperry designed ingenious experiments to test the abilities of split-brain patients when information was presented to either the right or left side of their brains. Because the connection between the hemispheres was severed, the two sides of the brain could not share what each side knew.

In 1981, Sperry was awarded the Nobel Prize in Physiology or Medicine for his research demonstrating the functional specialization of the right and left hemispheres of the brain and the role of the corpus callosum in transferring information between the hemispheres. Popular culture has exaggerated and overinterpreted Sperry's split-brain research, resulting in myths such as people having "left-brain" or "right-brain" personalities.

See also Corpus Callosum

Sphenopalatine Ganglioneuralgia

Ice cream headache or brain freeze. Some people can pronounce sphenopalatine ganglioneuralgia, but it's easier for them to just say "brain freeze" when they refer to the piercing head pain they get after drinking a milkshake too fast or taking a big bite of ice cream.

Brain freeze is caused by rapid cooling of blood vessels in the roof of the mouth. These blood vessels are surrounded by a network of nerves. When cold foods constrict the blood vessels, the nerves send pain signals to the brain. The body responds rapidly by trying to warm the cold area of the mouth by sending more blood. The rush of new blood expands the blood vessels that excite the nerves again. The constriction and expansion

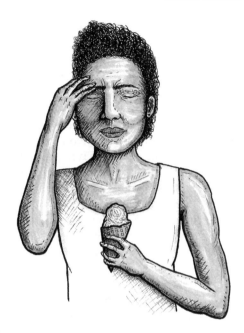

of cranial blood vessels caused by cold food may also result in the perception of pain.

Sphenopalatine ganglioneuralgia may sound like a severe neurological condition, but these headaches are fairly common, and while uncomfortable, they are temporary and not a cause for major concern. To avoid brain freeze, people can press their tongue to the roof of their mouth or just eat their frozen treats more slowly.

Spina Bifida
Neurodevelopmental birth defect that affects 1,500–2,000 babies each year in the United States. The phrase

spina bifida comes from the Latin words meaning "split spine."

Spina bifida occurs when the neural tube, the part of the growing fetus that develops into the brain and spinal cord, is not formed properly.[96] When the neural tube does not develop properly, three different types of spina bifida can occur. Spina bifida occulta is a mild form of spina bifida where one or more vertebrae are malformed, leaving a space between bones. Myelomeningocele is a severe form of spina bifida where the spinal cord, the meninges, or the spinal nerves extend out of an opening in the vertebrae. Meningocele is a rare form of spina bifida where the meninges surrounding the spinal cord extend out of an opening in the vertebrae. Babies born with a meningocele or a myelomeningocele have a small sac that protrudes out of their backs. If the spinal cord is not repaired, spina bifida can cause nerve damage, movement problems, bladder complications, and bowel difficulties.

Surgery performed on babies soon after they are born with spina bifida can often repair spinal cord damage. Also, with incredible skill and precision, pediatric neurosurgeons can perform surgery on babies who are still in the womb to limit the effects of spina bifida. Prenatal surgery can limit future damage to a baby's developing spinal cord and reduce the chance of movement problems as the baby grows.

See also Meninges; Spinal Cord; Vertebral Column

Spinal Cord

Connection between the brain and spinal nerves. Together with the brain, the spinal cord makes up the central nervous system. The adult human spinal cord is

44 to 46 cm (17 to 18 in) long and runs inside the bony vertebrae (backbone) within the vertebral canal.

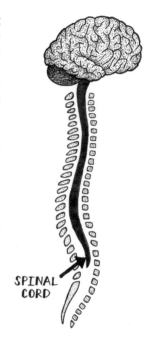

The spinal cord functions as a two-way street, sending information to and from the brain and the rest of the body. Axons travel the length of the spinal cord with signals from the brain to neurons that connect to muscles to control movement. Other axons travel in the spinal cord to bring sensory signals from the skin, joints, muscles, and some organs to the brain. Circuits of neurons within the spinal cord can also coordinate various reflexes without input from the brain.

The spinal cord contains neuronal cell bodies in addition to axonal pathways. When a cross section of the spinal cord is viewed, these cell bodies ("gray matter") form the shape of a butterfly. The upper sections of the butterfly's wings (the back of the spinal cord) are called the dorsal horns and contain neurons that send sensory information up to the brain. The lower sections of the butterfly's wings (the front of the spinal cord) are called the ventral horns and contain neurons that receive information from the brain to control muscles.

See also Axon; Neuron; Reflex; Vertebral Column

Stroke

Interruption of the blood supply to the brain, causing neurons to die. The brain receives a steady supply of oxygen, carbohydrates, amino acids, fats, hormones, and vitamins through its blood supply. But if the blood supply to the brain is stopped, then neurons will die.

Strokes happen when a blood vessel is blocked (ischemic stroke) or if a blood vessel bursts (hemorrhagic stroke). For example, a blood clot or a narrowing of an artery that brings blood to the brain can stop the flow of nutrients to neurons. Alternatively, blood vessels can bleed and cause strokes when there is swelling of a blood vessel wall or an abnormal connection between arteries and veins. Ischemic strokes are more common than hemorrhagic strokes.

Everyone should know how to recognize the initial signs of someone suffering from a stroke and what to do when this happens. An easy way to remember these signs is **FAST**: **F**ace drooping—**A**rm weakness—**S**peech problems—**T**ime (call 911). The effects of a stroke depend on the part of the brain that is damaged and the size of the injury. Paralysis on one side of the body is a common feature of a stroke, but cognitive issues involving memory, learning, attention, language disorders, and emotional problems may also occur.

High blood pressure, cigarette smoking, heart disease, and diabetes all increase the risk that a person will have a stroke. Therefore, lifestyle and behavioral changes can help reduce the risk of suffering a stroke. However, if someone does have a stroke, drugs that dissolve or remove the blood clot or stop bleeding may help minimize brain damage.

In the United States, approximately 795,000 people suffer a stroke and 140,000 people die from a stroke every year. Among the many people who have suffered from strokes, several are former US presidents, including John Quincy Adams (1767–1848), John Tyler (1790–1862), Millard Fillmore (1800–1974), Andrew Johnson (1808–1875), Chester Arthur (1829–1886), Woodrow Wilson (1856–1924), Warren G. Harding (1865–1923), Franklin D. Roosevelt (1882–1945), Dwight D. Eisenhower (1890–1969), Richard Nixon (1913–1994), and Gerald Ford (1913–2006).[97]

See also Circle of Willis

Synapse

Functional connection between one neuron and another neuron, muscle cell, or gland. Neurons are unlike any other cells in the body because they are capable of generating and sending electrical signals to other neurons, muscles, and organs. This communication between a neuron and another cell takes place at a synapse.

A synapse consists of a presynaptic ending, a postsynaptic ending, and a synaptic cleft. The majority of synapses are chemical in nature, consisting of a presynaptic ending of a neuron containing neurotransmitters, a postsynaptic ending as part of the receiving cell that

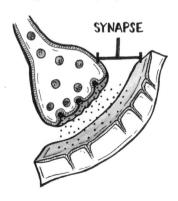

contains receptor sites for neurotransmitters, and the synaptic cleft—a small space (20–40 nanometers) between the presynaptic and postsynaptic endings. The nervous system also has electrical synapses where the presynaptic and postsynaptic endings are separated by about only 2 nanometers. At electrical synapses, small channels called gap junctions allow electric current to flow from one neuron to the next.

Textbooks often depict a synapse as a connection between the axon terminal of one neuron and a dendrite of another neuron (axodendritic synapse). However, synapses can also be formed by an axon terminal of one neuron and the cell body (axosomatic synapse) or axon (axoaxonic synapse) of another neuron.

In 1897, neuroscientist Charles S. Sherrington (1857–1952) coined the word *synapse* by combining the Greek words *syn* and *haptein*, meaning "together" and "to clasp." The word *synapse* was introduced in the book *A Textbook of Physiology, Part III: The Central Nervous System* (1897), written by Michael Foster (1836–1907) and assisted by Sherrington.

See also Action Potential; Neuron; Neurotransmitters; Sherrington, Charles Scott

Synesthesia

Perception of one sense that is activated simultaneously by another sense. Imagine that the sound of middle C looks green or that the color blue tastes sweet. Such examples of mixed sensory perceptions are what people with synesthesia (synesthetes) experience. In fact, the word *synesthesia* is derived from the Greek words *syn* and *aesthesis*, meaning "together" and "perception," respectively. Synesthesia is not a disease but rather a normal condition when the senses mix. The number of people with synesthesia may be as high as one in two hundred. Many people with synesthesia don't realize that other people do not experience the world the same way they do.

The most common form of synesthesia occurs when letters or numbers appear as specific colors, but any combination of senses can occur. Regardless of which senses mix, the perception is automatic, involuntary, and unique to each person. The perception is also the same every time it occurs.

Most synesthetes consider their abilities a gift and cannot imagine life without such perceptions. Many synesthetes use their perceptual abilities to drive creative musical and artistic pursuits and improve their memories. Well-known people who have said they have synesthesia include singers Tori Amos (1963–), Billie Eilish (2001–), Lorde (1996–), Billy Joel (1949–), Charli XCX (1992–), and Pharrell Williams (1973–); composer/pianist Duke Ellington (1899–1974); violinist Itzhak Perlman (1945–); and painter David Hockney (1937–).

Temporal Lobe

Part of each cerebral hemisphere located behind the frontal lobe and below the parietal lobe; responsible for encoding memories, processing information about hearing, and language comprehension. The location of the primary auditory cortex within the temporal lobe highlights the importance of this brain region in its role for hearing. Behind the auditory cortex is Wernicke's area, a region of cortex important for language comprehension. Other parts of the temporal lobe are involved with identification of visual objects and facial recognition.

The hippocampus is located in the temporal lobes; along with other parts of the cerebral cortex, such as the perirhinal, parahippocampal, and entorhinal cortical regions, it is key to memory formation and memory storage.

See also Hippocampus; Milner, Brenda; Molaison, Henry

Tourette Syndrome

Neurological disorder characterized by tics; involuntary, repetitive movements; and vocalizations. In 1825, French physician Jean-Marc Gaspard Itard (1774–1838) cared for a woman whose vocal tics had begun when she was seven years old. In 1885, another French physician, Georges Albert Edouard Brutus Gilles de la Tourette (1857–1904), published a report that described the symptoms of this patient and several other people with tics. In honor of this work, the syndrome was named after Gilles de la Tourette.

Tourette syndrome first appears when people are children, and the symptoms can last into adulthood. Common characteristics of Tourette syndrome are facial tics, such as eye blinks; vocal tics, such as sniffing, grunting, or throat clearing; and face, neck, and limb movements. To be diagnosed with Tourette syndrome, a person must have movement and vocal tics for at least a year and the tics must have started before the age of eighteen years. People with Tourette syndrome have normal intelligence and lead full lives.

The cause of Tourette syndrome is not known, but research indicates that there is likely a genetic component to the disorder. Sometimes tics cause only mild symptoms and medication is not needed. When tics interfere with daily activities, neuroleptic drugs such as haloperidol and pimozide that block the action of the neurotransmitter dopamine can lessen the symptoms of Tourette syndrome.

Several well-known professional athletes, including soccer player Tim Howard (1979–), baseball player Jim Eisenreich (1959–), NASCAR driver Steve Wallace (1987–), and basketball player Mahmoud Abdul-Rauf (1969–), have disclosed that they have Tourette syndrome.

See also Dopamine

Trepanation

Surgical procedure that scrapes or drills a hole into the skull. Trepanation is likely the earliest neurosurgical procedure ever performed and has been used since the late Paleolithic era. Archaeological evidence suggests that trepanation was practiced in ancient Egypt, Africa, India, China, the Americas, Rome, and Greece.[98] Today, trepanation (craniotomy) is used to treat head injuries, brain tumors, brain bleeding, and increased intracranial pressure.

Early trepans (also called trephines) made from hard stone or metal were used to scrape, drill, cut, or bore holes in the skull. Although the sizes of the holes found in trepanned skulls vary, some are several inches in diameter. Many skulls bearing trepanned holes show evidence that new bone grew after the surgery was conducted.

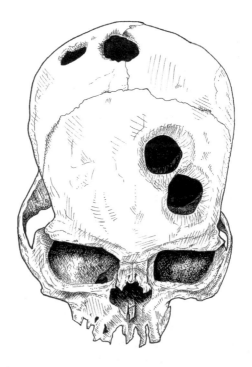

This indicates that the procedure was done on people who survived the gruesome practice: anesthetics were not discovered until thousands of years later.

Ancient people from around the world may have drilled skulls for religious rituals or for therapy. Trepanation was likely used to treat epilepsy, headaches, and psychiatric disorders. Perhaps people believed that evil spirits caused these ailments and reasoned that patients would be cured if these malevolent fiends escaped through a hole in the skull.

Interest in the nonmedical use of trepanning appeared in the late 1960s by a small group of people who believed that having a hole in their skull would increase brain blood flow, improve creativity and concentration, and heighten consciousness. There is no scientific evidence to support the belief that a hole in one's head provides improved cognitive abilities. Do-it-yourself trepanning carries a significant risk of infection and damage to the brain. Self-trepanning definitely qualifies as a "don't try this at home" activity. All trepanning should be left in the hands of a skilled neurosurgeon.

Twarog, Betty Mack (1927–2013)

American neuroscientist. Twarog graduated with an undergraduate degree in mathematics from Swarthmore College in 1948, but turned her attention to neurophysiology and neuropharmacology as a graduate student at Harvard University.

In the 1930s, scientists knew that cells in the gastrointestinal tract secreted substances that caused contraction of muscles in the intestines and that some chemicals could also make blood vessels contract. However, these early experiments did not identify the special chemical. Later chemical analysis of blood revealed that this chemical was serotonin (5-hydroxytryptamine). Twarog first showed that serotonin had the ability to affect the contraction of muscles in invertebrates such as mussels (*Mytilus edulis*); then to the surprise of many researchers, Twarog showed in 1953 that serotonin was located in the brains of rats, rabbits, and dogs. This was the first evidence that serotonin was found in the brains of vertebrate animals.[99]

Twarog's discovery of serotonin in the vertebrate brain suggested that the chemical might serve as a neurotransmitter. Her work set the foundation for research into antidepressant medications involving serotonin, such as the selective serotonin reuptake inhibitors fluoxetine (Prozac), paroxetine (Paxil, Pexeva), and sertraline (Zoloft).

See also Neurotransmitters; Serotonin

U mami

Taste of savory or meaty foods. Most people are familiar with four of the five basic tastes: sweet, sour, bitter, and salty. The fifth basic taste, umami, is a bit more difficult to appreciate. Umami is the flavor that gives beef and chicken its meaty taste.

Although the designation of umami as a basic taste was proposed by Japanese chemist Kikunae Ikeda (1864–1936) in 1908, it was not until the mid-1980s that umami was recognized alongside sour, sweet, bitter, and salty. Researchers have discovered that the taste of umami can be detected when molecules of glutamate bind to specific receptors on the tongue and in other areas of the mouth.[100] Umami receptors have also been found in the stomach, where signals are relayed to the brain to help with digestion.

The taste of umami can be found in cuisine that uses mushrooms, seaweed, parmesan cheese, anchovies, or soy sauce. The meaty, savory flavor of foods can be enhanced by adding monosodium glutamate to the dish.

Vagus Nerve

The tenth cranial nerve; the longest of the twelve cranial nerves. Think of the vagus nerve as a meandering pathway that sends information between the brain and organs in the chest and abdomen. In fact, the word *vagus* comes from the Latin word for "wandering" because this nerve connects the brain to so many different places in the body, including the heart, lungs, larynx, and digestive tract. The vagus nerve sends information from the brain to control the internal organs and from these organs back to the brain.

See also Cranial Nerves

Ventricles

Hollow series of connected spaces within the brain that are filled with cerebrospinal fluid. Like the famous canals of Venice, Italy, that help transport cargo and people throughout the waterways of the city, the ventricles of the brain provide a fluid passageway to move hormones, waste, and other materials throughout the central nervous system. While the Venetian canals are filled with water, the ventricles are filled with cerebrospinal fluid.

The lateral ventricles on each side of the brain are the largest of the ventricles. The lateral ventricle connects to the third ventricle via the interventricular foramen. The third ventricle then connects to the fourth ventricle through the cerebral aqueduct. The foramina of Luschka and the foramen of Magendie allow cerebrospinal fluid to flow out of the fourth ventricle into the space within the arachnoid.

In addition to providing a pathway to transport cerebrospinal fluid, the ventricles act as a safety mechanism in case of brain swelling. Head injury or infection may cause the brain to swell. Because the hard skull cannot expand to accommodate such swelling, pressure may damage brain tissue. The ventricles can expand slightly to lessen pressure and reduce the chance of brain injury.

See also Cerebrospinal Fluid; Meninges

Vertebral Column

Bones that make up the spinal column. The spinal cord of all vertebrate animals is encased in the bones of the spinal column (backbone) called vertebrae. Humans have thirty-three individual bones named for the region where they are located: seven cervical (neck) vertebrae; twelve thoracic (trunk) vertebrae, five lumbar (lower back) vertebrae, five sacral (pelvis) vertebrae, and four coccygeal (tailbone) vertebrae. In adults, the sacral and coccygeal vertebrae become fused to form the sacrum

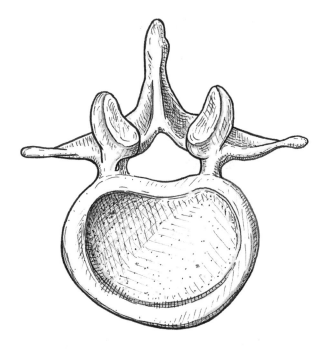

and coccyx, respectively. Each vertebra is separated by a disk that allows each bone to move slightly. The spinal cord runs through each vertebra within the vertebral foramen.

Most mammals—even a giraffe with a six-foot-long neck—have seven cervical vertebrae. Of course, there are exceptions to the rule: manatees and two-toed sloths have only six cervical vertebrae, ant bears have eight cervical vertebrae, and three-toed sloths have nine cervical vertebrae.

See also Spinal Cord

Vesalius, Andreas (1514–1564)

Flemish physician and anatomist. As a young medical student, Vesalius was schooled with the works of Greek physician Galen. Galen's teachings had dominated medical education for hundreds of years and were considered to be beyond reproach. But unlike Galen, Vesalius came about his anatomical knowledge by dissecting human

ANDREAS VESALIUS

bodies. Vesalius's study of human cadavers revealed that Galen had made significant errors about human anatomy. Vesalius described some of these errors in his 1543 manuscript titled *De humini corporus fabrica*, perhaps the greatest of Vesalius's published works.

De humini corporus fabrica is a collection of seven books with more than 270 detailed illustrations. Each book discusses a different part of the human body, with the fourth book concerned with nerves and the last book devoted to the brain and sensory organs. In contrast to many of his predecessors, Vesalius rejected the view that the ventricles were the source of cognition, and he showed that nerves did not branch from the heart. He also demonstrated that nerves were not hollow tubes as previously believed.

Not long after the publication of *De humini corporus fabrica*, Vesalius gave up his university appointment to serve as a court physician for Emperor Charles V (1500–1558) and later for King Philip II of Spain (1527–1598). After a pilgrimage to Jerusalem, Vesalius sailed to Italy in 1564. However, he became ill onboard and disembarked from a ship on the Greek island of Zakynthos (Zante), where he died days later.

Many people consider Vesalius to be the founder of modern medicine and the *De humini corporus fabrica* to be a scientific and artistic masterpiece.

See also Galen

Volkow, Nora (1956–)

Mexican American psychiatrist. Volkow was born in Mexico and received her medical degree in 1980 from the National University of Mexico in Mexico City. After medical school, Volkow continued her training by taking a psychiatric residency (1980–1984) at New York University.[101] As a researcher, Volkow used brain imaging to study how drugs such as alcohol, cocaine, and cannabinoids affect the human brain. She has also been a driving force to promote public awareness that drug addiction is a brain disease.

In 2003, Volkow was appointed as the director of the National Institute on Drug Abuse (NIDA), one of the twenty-seven institutes and centers at the National Institutes of Health. As director of the NIDA, Volkow oversees a budget of approximately $1.4 billion for drug abuse and addiction research, education, and training.

Volkow has been an active public spokesperson fighting substance abuse, explaining the consequences of addiction, and reducing the stigma attached to addiction. She was named as one of *Time* magazine's "Most Influential People in the World" in 2007 and was profiled by the *60 Minutes* television show in 2012. Volkow also has an interesting family history: she is the great-granddaughter of Leon Trotsky (1879–1940), the Russian Marxist revolutionary leader.

See also Alcohol; Cocaine; Marijuana

Volta, Alessandro (1745–1827)

Italian physicist who discovered the voltaic pile, a forerunner of the electric battery. The saying "Necessity is the mother of invention" certainly rang true for Alessandro Volta. Volta needed a way to challenge his scientific rival Luigi Galvani and show that animals did not have their own innate electricity. Volta believed that two different metals were responsible for the electricity and that the muscles were just responding to the electricity, not actually producing their own electricity as Galvani hypothesized. This scientific disagreement with Galvani provided Volta with the motivation to invent the voltaic pile.

In the late 1790s, Volta stacked alternating discs of zinc and copper separated by pieces of cloth or cardboard soaked in brine. This arrangement, called a voltaic pile, was capable of generating an adjustable, steady electric current. Volta used his invention in experiments to show that electric current did not need nerves or muscles.

In honor of Volta's accomplishments, the International Exposition of Electricity (International Electrical Congress) of 1881 officially named the unit of electric potential the "volt."

See also Galvani, Luigi

Wernicke, Carl (1848–1905)

German physician (psychiatrist/neurologist). Carl Wernicke continued the work of Paul Broca (1824–1880) to understand how speech is localized to specific areas of the brain. Born in Prussia (now Poland), Wernicke studied medicine at the University of Breslau and served as a psychiatrist in various hospitals.[102] After collecting histories of patients with language disorders, Wernicke published his findings in the book titled *The Aphasia Symptom Complex* (1874). In this book, Wernicke described a language disorder caused by damage to the superior temporal gyrus. People with this disorder, later called Wernicke's aphasia, have problems comprehending written and spoken language. People with Wernicke's aphasia can speak fluently, but the words have no meaning.

Wernicke advocated for a neurobiological classification of psychiatric disorders. Although he was not able to assign specific brain areas to certain mental capacities, Wernicke argued, in a view still held today, that higher cognition function is the result of interactions between regions with their own specific roles.

See also Broca, Paul

Zika Virus

Virus belonging to the family Flaviviridae, the type of virus responsible for dengue fever, chikungunya, and yellow fever. Zika virus can be spread to people through the bite of an infected *Aedes* species mosquito. Adults who are infected with the Zika virus often have only mild symptoms, such as a fever, rash, headache, conjunctivitis, and joint pain. Unfortunately, the virus can be transmitted during pregnancy from

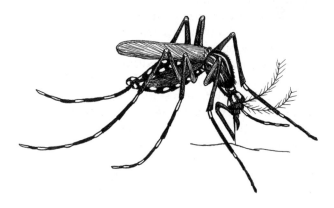

mothers to their infants. Infants infected with Zika virus may be born with microcephaly and other severe brain abnormalities.

The only way to diagnose Zika virus is by testing the blood, saliva, or urine of a sick person for the presence of the virus. Because there is no vaccine for Zika virus, the best strategy is to avoid the virus in the first place. For example, people visiting places where Zika virus is active should wear long-sleeved shirts and pants, use mosquito screens, and apply insect repellents to prevent mosquito bites.

The Zika virus was identified in humans about fifty years ago, and outbreaks of infections have occurred around the world through the decades. In 2015, a large number of Zika virus cases appeared in Brazil and then in other countries. However, since 2018, the US has had no cases of Zika virus transmitted by mosquitoes.

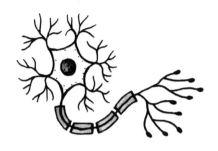

Useful Resources

Books

American Psychiatric Association. *Diagnostic and Statistical Manual of Mental Disorders*. 5th ed. Arlington, VA: American Psychiatric Association, 2013.

Catani, Marco, and Sandrone, Stefano. *Brain Renaissance: From Vesalius to Modern Neuroscience*. New York: Oxford University Press, 2015.

Chudler, Eric H., and Johnson, Lise A. *Brain Bytes: Quick Answers to Quirky Questions about the Brain*. New York: W. W. Norton, 2017.

Eagleman, David. *The Brain: The Story of You*. New York: Pantheon Books, 2015.

Finger, Stanley. *Minds behind the Brain: A History of the Pioneers and Their Discoveries*. New York: Oxford University Press, 2000.

Gross, Charles G. *A Hole in the Head: More Tales in the History of Neuroscience*. Cambridge, MA: MIT Press, 2009.

Kandel, Eric R., Koester, J. D., Mack, S. H., and Siegel-baum, S. A. *Principles of Neural Science*. 6th ed. New York: McGraw-Hill, 2021.

McComas, Alan J. *Galvani's Spark: The Story of the Nerve Impulse*. New York: Oxford University Press, 2011.

Websites

BrainFacts. Society for Neuroscience. https://www.brainfacts.org/.

Centers for Disease Control and Prevention. https://www.cdc.gov/.

Dana Foundation. https://dana.org/.

National Institute of Neurological Disorders and Stroke. National Institutes of Health. https://www.ninds.nih.gov/.

Neuroscience for Kids. http://faculty.washington.edu/chudler/neurok.html.

References

1 Ramón y Cajal, S., Charlas de café: Pensamientos, anécdotas y confidencias. 3rd ed. Madrid: Imprenta de Juan Pueyo Luna, 1922.

2 Mercante, G., Ferreli, F., De Virgilio, A., Gaino, F., Di Bari, M., Colombo, G., Russo, E., Costantino, A., Pirola, F., Cugini, G., Malvezzi, L., Morenghi, E., Azzolini, E., Lagioia, M., and Spriano, G., Prevalence of taste and smell dysfunction in coronavirus disease 2019. JAMA Otolaryngol Head Neck Surg. 146(8):723–28, 2020.

3 Wiens, F., Zitzmann, A., Lachance, M.A., Yegles, M., Pragst, F., Wurst, F.M., von Holst, D., Guan, S.L., and Spanagel, R., Chronic intake of fermented floral nectar by wild treeshrews. Proc Natl Acad Sci U S A. 105(30):10426–31, 2008.

4 Sarva, H., Deik, A., and Severt, W.L., Pathophysiology and treatment of alien hand syndrome. Tremor Other Hyperkinet Mov (NY). 4:241, 2014.

5 Iversen, L., Speed, Ecstasy, Ritalin: The Science of Amphetamines. Oxford, UK: Oxford University Press, 2008.

6 US Drug Enforcement Administration, Drug scheduling. September 2, 2021, https://www.dea.gov/drug-scheduling.

7 Zald, D.H., The human amygdala and the emotional evaluation of sensory stimuli. Brain Res Rev. 41(1):88–123, 2003.

8 Kandel, E.R., Small systems of neurons. Sci Am. 241(3):66–76, 1979.

9 Clarke, E., and Stannard, J., Aristotle on the anatomy of the brain. J Hist Med Allied Sci. 18:130–48, 1963.

10 Faraone, S.V., The pharmacology of amphetamine and methylphenidate: Relevance to the neurobiology of attention-deficit/hyperactivity disorder and other psychiatric comorbidities. Neurosci Biobehav Rev. 87:255–70, 2018.

11 Centers for Disease Control and Prevention, Data & statistics on autism spectrum disorder. August 30, 2021, https://www.cdc.gov/ncbddd/autism/data.html.

12 Treffert, D.A., The savant syndrome: An extraordinary condition. A synopsis: Past, present, future. Philos Trans R Soc Lond B Biol Sci. 364(1522):1351–7, 2009.

13 Zargaran, A., Mehdizadeh, A., Zarshenas, M.M., and Mohagheghzadeh, A., Avicenna (980–1037 AD). J Neurol. 259(2):389–90, 2012.

14 Scully, T., Neuroscience: The great squid hunt. Nature. 454(7207): 934–36, 2008.

15 Abbott, A., Documentary follows implosion of billion-euro brain project. Nature. 588(7837):215–16, 2020.

16 Koroshetz, W., Gordon, J., Adams, A., Beckel-Mitchener, A., Churchill, J., Farber, G., Freund, M., Gnadt, J., Hsu, N.S., Langhals, N., Lisanby, S., Liu, G., Peng, G.C.Y., Ramos, K., Steinmetz, M., Talley, E., and White, S., The state of the NIH BRAIN Initiative. J Neurosci. 38(29):6427–38, 2018.

17 Theil, S., Why the Human Brain Project went wrong—and how to fix it. Scientific American. 313:36–42, 2015.

18 Yong, E., The Human Brain Project hasn't lived up to its promise. Atlantic, July 22, 2019.

19 Domanski, C.W., Mysterious "Monsieur Leborgne": The mystery of the famous patient in the history of neuropsychology is explained. J Hist Neurosci. 22(1):47–52, 2013.

20 Czajkowski, N., Kendler, K.S., Torvik, F.A., Ystrom, E., Rosenstrom, T., Gillespie, N.A., and Reichborn-Kjennerud, T., Caffeine consumption, toxicity, tolerance and withdrawal; shared genetic influences with normative personality and personality disorder traits. Exp Clin Psychopharmacol. 29(6):650–58, 2021.

21 Edelstyn, N.M., and Oyebode, F., A review of the phenomenology and cognitive neuropsychological origins of the Capgras syndrome. Int J Geriatr Psychiatry. 14(1):48–59, 1999.

22 Miwa, H., and Mizuno, Y., Capgras syndrome in Parkinson's disease. J Neurol. 248(9):804–5, 2001.

23 Josephs, K.A., Capgras syndrome and its relationship to neurodegenerative disease. Arch Neurol. 64(12):1762–66, 2007.

24 Van Essen, D.C., Donahue, C.J., and Glasser, M.F., Development and evolution of cerebral and cerebellar cortex. Brain Behav Evol. 91(3):158–69, 2018.

25 Yu, F., Jiang, Q.J., Sun, X.Y., and Zhang, R.W., A new case of complete primary cerebellar agenesis: Clinical and imaging findings in a living patient. Brain. 138(Pt 6):e353, 2015.

26 World Health Organization, Deafness and hearing loss. April 1, 2021, https://www.who.int/news-room/fact-sheets/detail/deafness-and-hearing-loss.

27 Rose, N., and Abi-Rached, J.M., Neuro: The New Brain Sciences and the Management of the Mind. Princeton, NJ: Princeton University Press, 2013.

28 Scorza, K.A., and Cole, W., Current concepts in concussion: Initial evaluation and management. Am Fam Physician. 99(7):426–34, 2019.

29 Shields, S.D., Deng, L., Reese, R.M., Dourado, M., Tao, J., Foreman, O., Chang, J.H., and Hackos, D.H., Insensitivity to pain upon adult-onset deletion of Nav1.7 or its blockade with selective inhibitors. J Neurosci. 38(47):10180–201, 2018.

30 Song, E., Zhang, C., Israelow, B., Lu-Culligan, A., Prado, A.V., Skriabine, S., Lu, P., Weizman, O.E., Liu, F., Dai, Y., Szigeti-Buck, K., Yasumoto, Y., Wang, G., Castaldi, C., Heltke, J., Ng, E., Wheeler, J., Alfajaro, M.M., Levavasseur, E., Fontes, B., Ravindra, N.G., Van Dijk, D., Mane, S., Gunel, M., Ring, A., Kazmi, S.A.J., Zhang, K., Wilen, C.B., Horvath, T.L., Plu, I., Haik, S., Thomas, J.L., Louvi, A., Farhadian, S.F., Huttner, A., Seilhean, D., Renier, N., Bilguvar, K., and Iwasaki, A., Neuroinvasion of SARS-CoV-2 in human and mouse brain. J Exp Med. 218(3):e20202135, 2021.

31 Andrabi, M.S., and Andrabi, S.A., Neuronal and cerebrovascular complications in coronavirus disease 2019. Front Pharmacol. 11:570031, 2020.

32 Mancuso, L., Uddin, L.Q., Nani, A., Costa, T., and Cauda, F., Brain functional connectivity in individuals with callosotomy and agenesis of the corpus callosum: A systematic review. Neurosci Biobehav Rev. 105:231–48, 2019.

33 Suarez, R., Paolino, A., Fenlon, L.R., Morcom, L.R., Kozulin, P., Kurniawan, N.D., and Richards, L.J., A pan-mammalian map of interhemispheric brain connections predates the evolution of the corpus callosum. Proc Natl Acad Sci U S A. 115(38):9622–27, 2018.

34 Debruyne, H., Portzky, M., Van den Eynde, F., and Audenaert, K., Cotard's syndrome: A review. Curr Psychiatry Rep. 11(3):197–202, 2009.

35 Berrios, G.E., and Luque, R., Cotard's delusion or syndrome? A conceptual history. Compr Psychiatry. 36(3):218–23, 1995.

36 Klein, M.O., Battagello, D.S., Cardoso, A.R., Hauser, D.N., Bittencourt, J.C., and Correa, R.G., Dopamine: Functions, signaling, and association with neurological diseases. Cell Mol Neurobiol. 39(1):31–59, 2019.

37 Brisch, R., Saniotis, A., Wolf, R., Bielau, H., Bernstein, H.G., Steiner, J., Bogerts, B., Braun, K., Jankowski, Z., Kumaratilake, J., Henneberg, M., and Gos, T., The role of dopamine in schizophrenia from a neurobiological and evolutionary perspective: Old fashioned, but still in vogue. Front Psychiatry. 5:47, 2014.

38 Sacks, O., Awakenings. London: Duckworth, 1973.

39 Breasted, H., The Edwin Smith Surgical Papyrus. Chicago: University of Chicago Press, 1930.

40 Diamond, M.C., Scheibel, A.B., Murphy, G.M., Jr., and Harvey, T., On the brain of a scientist: Albert Einstein. Exp Neurol. 88(1):198–204, 1985.

41 Anderson, B., and Harvey, T., Alterations in cortical thickness and neuronal density in the frontal cortex of Albert Einstein. Neurosci Lett. 210(3):161–64, 1996.

42 Colombo, J.A., Reisin, H.D., Miguel-Hidalgo, J.J., and Rajkowska, G., Cerebral cortex astroglia and the brain of a genius: A propos of A. Einstein's. Brain Res Rev. 52(2):257–63, 2006.

43 Falk, D., Lepore, F.E., and Noe, A., The cerebral cortex of Albert Einstein: A description and preliminary analysis of unpublished photographs. Brain. 136(Pt 4):1304–27, 2013.

44 Atta, K., Forlenza, N., Gujski, M., Hashmi, S., and Isaac, G., Delusional misidentification syndromes: Separate disorders or unusual presentations of existing DSM-IV categories? Psychiatry (Edgmont). 3(9):56–61, 2006.

45 O'Brien, G., Rosemary Kennedy: The importance of a historical footnote. J Fam Hist. 29(3):225–36, 2004.

46 Zhang, W., Xiong, B.R., Zhang, L.Q., Huang, X., Yuan, X., Tian, Y.K., and Tian, X.B., The role of the GABAergic system in diseases of the central nervous system. Neuroscience. 470:88–99, 2021.

47 Macmillan, M., and Lena, M.L., Rehabilitating Phineas Gage. Neuropsychol Rehabil. 20(5):641–58, 2010.

48 Piccolino, M., and Bresadola, M., Shocking Frogs: Galvani, Volta, and the Electric Origins of Neuroscience. Oxford, UK: Oxford University Press, 2013.

49 Bresadola, M., Animal electricity at the end of the eighteenth century: The many facets of a great scientific controversy. J Hist Neurosci. 17(1):8–32, 2008.

50 Butt, A., and Verkhratsky, A., Neuroglia: Realising their true potential. Brain Neurosci Adv. 2:2398212818817495, 2018.

51 van Gijn, J., Camillo Golgi (1843–1926). J Neurol. 248(6):541–42, 2001.

52 Bentivoglio, M., Cotrufo, T., Ferrari, S., Tesoriero, C., Mariotto, S., Bertini, G., Berzero, A., and Mazzarello, P., The original histological slides of Camillo Golgi and his discoveries on neuronal structure. Front Neuroanat. 13:3, 2019.

53 Bir, S.C., Ambekar, S., Kukreja, S., and Nanda, A., Julius Caesar Arantius (Giulio Cesare Aranzi, 1530–1589) and the hippocampus of the human brain: History behind the discovery. J Neurosurg. 122(4):971–75, 2015.

54 Maguire, E.A., Woollett, K., and Spiers, H.J., London taxi drivers and bus drivers: A structural MRI and neuropsychological analysis. Hippocampus. 16(12):1091–101, 2006.

55 Azam, S., Haque, M.E., Balakrishnan, R., Kim, I.S., and Choi, D.K., The ageing brain: Molecular and cellular basis of neurodegeneration. Front Cell Dev Biol. 9:683459, 2021.

56 Arevalo, J., Wojcieszek, J., and Conneally, P.M., Tracing Woody Guthrie and Huntington's disease. Semin Neurol. 21(2):209–23, 2001.

57 US Department of Health and Human Services, Office of Healthy Homes and Lead Hazard Controls. American Healthy Homes Survey: Lead and Arsenic Findings. Washington, DC: US Department of Health and Human Services, 2011.

58 Mason, L.H., Harp, J.P., and Han, D.Y., Pb neurotoxicity: Neuropsychological effects of lead toxicity. Biomed Res Int. 2014:840547, 2014.

59 Sandrone, S., Rita Levi-Montalcini (1909–2012). J Neurol. 260(3): 940–41, 2013.

60 Rodrigues e Silva, A.M., Geldsetzer, F., Holdorff, B., Kielhorn, F.W., Balzer-Geldsetzer, M., Oertel, W.H., Hurtig, H., and Dodel, R., Who was the man who discovered the "Lewy bodies"? Mov Disord. 25(12):1765–73, 2010.

61 Loewi, O., From the Workshop of Discoveries. Lawrence: University of Kansas Press, 1953.

62 Hofmann, A., LSD—My Problem Child. New York: McGraw-Hill, 1980.

63 US DEA, Drug scheduling. https://www.dea.gov/drug-scheduling.

64 De Gregorio, D., Aguilar-Valles, A., Preller, K.H., Heifets, B.D., Hibicke, M., Mitchell, J., and Gobbi, G., Hallucinogens in mental health: Preclinical and clinical studies on LSD, psilocybin, MDMA, and ketamine. J Neurosci. 41(5):891–900, 2021.

65 Garcia-Romeu, A., Kersgaard, B., and Addy, P.H., Clinical applications of hallucinogens: A review. Exp Clin Psychopharmacol. 24(4):229–68, 2016.

66 National Institute on Drug Abuse, Is marijuana addictive. https://www.drugabuse.gov/publications/research-reports/marijuana/marijuana-addictive.

67 Hirschhorn, N., Feldman, R.G., and Greaves, I.A., Abraham Lincoln's blue pills: Did our 16th president suffer from mercury poisoning? Perspect Biol Med. 44(3):315–32, 2001.

68 Corkin, S., Lasting consequences of bilateral medial temporal lobectomy: Clinical course and experimental findings in H. M. Seminars in Neurology. 4:249–59, 1984.

69 Squire, L.R., The legacy of patient H. M. for neuroscience. Neuron. 61(1):6–9, 2009.

70 Annese, J., Schenker-Ahmed, N.M., Bartsch, H., Maechler, P., Sheh, C., Thomas, N., Kayano, J., Ghatan, A., Bresler, N., Frosch,

M.P., Klaming, R., and Corkin, S., Postmortem examination of patient H. M.'s brain based on histological sectioning and digital 3D reconstruction. Nat Commun. 5:3122, 2014.

71 The Brain Observatory. May 1, 2021, https://www.thebrain observatory.org/.

72 US DEA, Drug scheduling. https://www.dea.gov/drug-scheduling.

73 Kimmel, J., What it feels like to have narcolepsy. Esquire, August 2003, 74.

74 Organization for the Prohibition of Chemical Weapons, Chemical Weapons Convention. https://www.opcw.org/chemical-weapons -convention.

75 Wexler, A., and Reiner, P.B., Oversight of direct-to-consumer neu-rotechnologies. Science. 363(6424):234–35, 2019.

76 Azevedo, F.A., Carvalho, L.R., Grinberg, L.T., Farfel, J.M., Ferretti, R.E., Leite, R.E., Jacob Filho, W., Lent, R., and Herculano-Houzel, S., Equal numbers of neuronal and nonneuronal cells make the human brain an isometrically scaled-up primate brain. J Comp Neurol. 513(5):532–41, 2009.

77 Voss, P., Thomas, M.E., Cisneros-Franco, J.M., and de Villers-Sidani, E., Dynamic brains and the changing rules of neuroplasticity: Implications for learning and recovery. Front Psychol. 8:1657, 2017.

78 Forbes, S.C., Holroyd-Leduc, J.M., Poulin, M.J., and Hogan, D.B., Effect of nutrients, dietary supplements and vitamins on cognition: A systematic review and meta-analysis of randomized controlled trials. Can Geriatr J. 18(4):231–45, 2015.

79 Boyden, E.S., A history of optogenetics: The development of tools for controlling brain circuits with light. F1000 Biol Rep. 3:11, 2011.

80 Sim, J.H., Roosli, C., Chatzimichalis, M., Eiber, A., and Huber, A.M., Characterization of stapes anatomy: Investigation of human and guinea pig. J Assoc Res Otolaryngol. 14(2):159–73, 2013.

81 Parkinson, J., An Essay on the Shaking Palsy. London: Sherwood Neely and Jones, 1817.

82 Feindel, W., The physiologist and the neurosurgeon: The enduring influence of Charles Sherrington on the career of Wilder Penfield. Brain. 130(Pt 11):2758–65, 2007.

83 Corrow, S.L., Dalrymple, K.A., and Barton, J.J., Prosopagnosia: Current perspectives. Eye Brain. 8:165–75, 2016.

84 Centers for Disease Control and Prevention, Rabies. April 2, 2020, https://www.cdc.gov/rabies/index.html.

85 Rapport, R., Nerve Endings: The Discovery of the Synapse. New York: W. W. Norton, 2005.

86 Sotelo, C., Viewing the brain through the master hand of Ramón y Cajal. Nat Rev Neurosci. 4(1):71–77, 2003.

87 Ramón y Cajal, S., Charlas de café: Pensamientos, anécdotas y confidencias. 3rd ed. Madrid: Imprenta de Juan Pueyo Luna, 1922.

88 Berciano, J., and Lafarga, M., Pioneers in neurology: Santiago Ramón y Cajal (1852–1934). J Neurol. 248(2):152–53, 2001.

89 Liu, N., Xiao, Y., Zhang, W., Tang, B., Zeng, J., Hu, N., Chandan, S., Gong, Q., and Lui, S., Characteristics of gray matter alterations in never-treated and treated chronic schizophrenia patients. Transl Psychiatry. 10(1):136, 2020.

90 Ehrlich, S., Geisler, D., Yendiki, A., Panneck, P., Roessner, V., Calhoun, V.D., Magnotta, V.A., Gollub, R.L., and White, T., Associations of white matter integrity and cortical thickness in patients with schizophrenia and healthy controls. Schizophr Bull. 40(3):665–74, 2014.

91 Fusar-Poli, P., and Politi, P., Paul Eugen Bleuler and the birth of schizophrenia (1908). Am J Psychiatry. 165(11):1407, 2008.

92 Gershon, M.D., and Tack, J., The serotonin signaling system: From basic understanding to drug development for functional GI disorders. Gastroenterology. 132(1):397–414, 2007.

93 Weisel-Eichler, A., and Libersat, F., Venom effects on monoaminergic systems. J Comp Physiol A Neuroethol Sens Neural Behav Physiol. 190(9):683–90, 2004.

94 Breathnach, C.S., Charles Scott Sherrington (1857–1952). J Neurol. 252(8):1000–1001, 2005.

95 Damasio, A.R., Reflecting on the work of R. W. Sperry. Trends in Neurosciences. 5:222–24, 1982.

96 Centers for Disease Control and Prevention, Spina bifida. https://www.cdc.gov/ncbddd/spinabifida/index.html.

97 Jones, J.M., and Jones, J.L., Presidential stroke: United States presidents and cerebrovascular disease. CNS Spectr. 11(9):674–78, 719, 2006.

98 Bynum, B., and Bynum, H., Trepanned cranium. Lancet. 392(10142): 112, 2018.

99 Twarog, B.M., Serotonin: History of a discovery. Comp Biochem
    Physiol C Comp Pharmacol Toxicol. 91(1):21–24, 1988.

100 Hartley, I.E., Liem, D.G., and Keast, R., Umami as an "alimen-
    tary" taste: A new perspective on taste classification. Nutrients.
    11(1):182, 2019.

101 National Institute on Drug Abuse, Biography of Dr. Nora Vol-
    kow. March 12, 2021, https://www.drugabuse.gov/about-nida
    /directors-page/biography-dr-nora-volkow.

102 Pillmann, F., Carl Wernicke (1848–1905). J Neurol. 250(11):1390–
    91, 2003.

Collect Them All

Fungipedia

A Brief Compendium of Mushroom Lore — Lawrence Millman

Florapedia

A Brief Compendium of Floral Lore — Carol Gracie

Birdpedia

A Brief Compendium of Avian Lore — Christopher W. Leahy

Treepedia

A Brief Compendium of Arboreal Lore — Joan Maloof

Dinopedia

A Brief Compendium of Dinosaur Lore — Darren Naish

Insectpedia

A Brief Compendium of Insect Lore — Eric R. Eaton

Geopedia

A Brief Compendium of Geologic Curiosities — Marcia Bjornerud

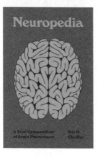

Neuropedia

A Brief Compendium of Brain Phenomena — Eric H. Chudler

Discover the rest of the series